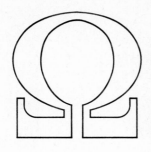 Perspectives in Mathematical Logic

Ω-Group:
R. O. Gandy, H. Hermes, A. Levy, G. H. Müller,
G. E. Sacks, D. S. Scott

Marian B. Pour-El
J. Ian Richards

Computability in Analysis and Physics

Springer-Verlag
Berlin Heidelberg New York
London Paris Tokyo

Marian B. Pour-El
J. Ian Richards
School of Mathematics, University of Minnesota,
Minneapolis, MN 55455, USA

In 1987, editorial responsibility for this series was assumed by the Association for Sympolic Logic. The current members of the editorial board are: S. Fefermann, M. Lerman (managing editor), A. Levy, A. J. MacIntyre, Y. N. Moschovakis, G. H. Müller

AMS Subject Classification (1980):
03D80, 03F60, 46N05

ISBN 3-540-50035-9 Springer-Verlag Berlin Heidelberg NewYork
ISBN 0-387-50035-9 Springer-Verlag NewYork Berlin Heidelberg

Library of Congress Cataloging-in-Publication Data
Pour-El, Marian B. (Marian Boykan), 1928-
Computability in analysis and physics / Marian B. Pour-El,
J. Ian Richards.
(Perspectives in mathematical logic)
Bibliography: p. Includes index.
ISBN 0-387-50035-9 (U.S.)
1. Computable functions.   2. Mathematical analysis.   3. Physics.
I. Richards, J. Ian (Jonathan Ian, 1936-.   II. Title.
III. Series.
QA9.59.P68 1988

This work is subject to copyright. All rights are reserved, whether the whole or part of the material is concerned, specifically the rights of translation, reprinting, reuse of illustrations, recitation, broadcasting, reproduction on microfilms or in other ways, and storage in data banks. Duplication of this publication or parts thereof is only permitted under the provisions of the German Copyright Law of September 9, 1965, in its version of June, 24, 1985, and a copyright fee must always be paid. Violations fall under the prosecution act of the German Copyright Law.

© Springer-Verlag, Berlin Heidelberg 1989
Printed in Germany

The use of registered names, trademarks, etc. in this publication does not imply, even in the absence of a specific statement, that such names are exempt from the relevant protective laws and regulations and therefore free for general use.

Typesetting: Asco Trade Typesetting Ltd., Hong Kong;
Offsetprinting: Mercedes-Druck GmbH, Berlin; Binding: Lüderitz & Bauer, Berlin
2141/3020-543210   Printed on acid-free paper

Preface to the Series
Perspectives in Mathematical Logic
(Edited by the Ω-group for "Mathematische Logik" of the
Heidelberger Akademie der Wissenschaften)

On Perspectives. *Mathematical logic arose from a concern with the nature and the limits of rational or mathematical thought, and from a desire to systematise the modes of its expression. The pioneering investigations were diverse and largely autonomous. As time passed, and more particularly in the last two decades, interconnections between different lines of research and links with other branches of mathematics proliferated. The subject is now both rich and varied. It is the aim of the series to provide, as it were, maps of guides to this complex terrain. We shall not aim at encyclopaedic coverage; nor do we wish to prescribe, like Euclid, a definitive version of the elements of the subject. We are not committed to any particular philosophical programme. Nevertheless we have tried by critical discussion to ensure that each book represents a coherent line of thought; and that, by developing certain themes, it will be of greater interest than a mere assemblage of results and techniques.*

The books in the series differ in level: some are introductory, some highly specialised. They also differ in scope: some offer a wide view of an area, others present a single line of thought. Each book is, at its own level, reasonably self-contained. Although no book depends on another as prerequisite, we have encouraged authors to fit their book in with other planned volumes, sometimes deliberately seeking coverage of the same material from different points of view. We have tried to attain a reasonable degree of uniformity of notation and arrangement. However, the books in the series are written by individual authors, not by the group. Plans for books are discussed and argued about at length. Later, encouragement is given and revisions suggested. But it is the authors who do the work; if, as we hope, the series proves of value, the credit will be theirs.

History of the Ω-Group. *During 1968 the idea of an integrated series of monographs on mathematical logic was first mooted. Various discussions led to a meeting at Oberwolfach in the spring of 1969. Here the founding members of the group (R.O. Gandy, A. Levy, G.H. Müller, G. Sacks, D.S. Scott) discussed the project in earnest and decided to go ahead with it. Professor F.K. Schmidt and Professor Hans Hermes gave us encouragement and support. Later Hans Hermes joined the group. To begin with all was fluid. How ambitious should we be? Should we write the books ourselves? How long would it take? Plans for authorless books were promoted, savaged and scrapped. Gradually there emerged a form and a method. At the end of an infinite discussion we found our name, and that of the series. We established our centre in Heidelberg. We agreed to meet twice a year together with authors, consultants and assistants, generally in Oberwolfach. We soon found the value of collaboration: on the one hand the permanence of the founding group gave coherence to the overall plans; on the other hand the stimulus of new contributors kept the project alive and flexible. Above all, we found how intensive discussion could modify the authors' ideas and our own. Often the battle ended with a detailed plan for a better book which the author was keen to write and which would indeed contribute a perspective.*

Oberwolfach, September 1975

Acknowledgements. *In starting our enterprise we essentially were relying on the personal confidence and understanding of Professor Martin Barner of the Mathematisches Forschungsinstitut Oberwolfach, Dr. Klaus Peters of Springer-Verlag and Dipl.-Ing. Penschuck of the Stiftung Volkswagenwerk. Through the Stiftung Volkswagenwerk we received a generous grant (1970–1973) as an initial help which made our existence as a working group possible.*

Since 1974 the Heidelberger Akademie der Wissenschaften (Mathematisch-Naturwissenschaftliche Klasse) has incorporated our enterprise into its general scientific program. The initiative for this step was taken by the late Professor F.K. Schmidt, and the former President of the Academy, Professor W. Doerr.

Through all the years, the Academy has supported our research project, especially our meetings and the continuous work on the Logic Bibliography, in an outstandingly generous way. We could always rely on their readiness to provide help wherever it was needed.

Assistance in many various respects was provided by Drs. U. Felgner and K. Gloede (till 1975) and Drs. D. Schmidt and H. Zeitler (till 1979). Last but not least, our indefatigable secretary Elfriede Ihrig was and is essential in running our enterprise.

We thank all those concerned.

Heidelberg, September 1982

R.O. Gandy   H. Hermes
A. Levy   G.H. Müller
G. Sacks   D.S. Scott

Authors' Preface

This book is concerned with the computability or noncomputability of standard processes in analysis and physics.

Part I is introductory. It provides the basic prerequisites for reading research papers in computable analysis (at least when the reasoning is classical). The core of the book is contained in Parts II and III. Great care has been taken to motivate the material adequately. In this regard, the introduction to the book and the introductions to the chapters play a major role. We suggest that the reader who wishes to skim begin with the introduction to the book and then read the introductions to Chapters 2, 3, and 4—with a brief glance at the introductions to Chapters 0 and 1.

The book is written for a mixed audience. Although it is intended primarily for logicians and analysts, it should be of interest to physicists and computer scientists—in fact to graduate students in any of these fields. The work is self-contained. Beyond the section on Prerequisites in Logic and Analysis, no further background is necessary. The reasoning used is classical—i.e. in the tradition of classical mathematics. Thus it is not intuitionist or constructivist in the sense of Brouwer or Bishop.

It is a pleasure to thank our mathematical colleagues for stimulating conversations as well as helpful comments and suggestions: Professors Oliver Aberth, John Baldwin, William Craig, Solomon Feferman, Alexander Kechris, Manuel Lerman, Saunders MacLane, Yiannis Moschovakis, Jack Silver, Dana Scott and Pat Suppes. Thanks are also due to Professor Yoshi Oono of the Department of Physics of the University of Illinois in Urbana, who, together with some of his colleagues, went through the manuscript in a seminar and commented extensively on questions of interest to physicists.

The Mathematical Sciences Research Institute at Berkeley provided hospitality and support to one of the authors (M.B.P.) during the fall of 1985.

Kate Houser did a superb job of typing the manuscript.

Finally, we wish to thank Professor Gert Müller and Springer-Verlag for their help above and beyond the call of duty.

We hope that this short monograph will provide an easy entrance into research in this area.

January 1988
Minneapolis, Minnesota

Marian Boykan Pour-El
Ian Richards

Contents

Introduction 1
Prerequisites from Logic and Analysis 6

Part I. Computability in Classical Analysis 9

Chapter 0. An Introduction to Computable Analysis 11
    Introduction 11

    1. Computable Real Numbers 13
    2. Computable Sequences of Real Numbers 17
    3. Computable Functions of One or Several Real Variables . . . 24
    4. Preliminary Constructs in Analysis 28
    5. Basic Constructs of Analysis 33
    6. The Max-Min Theorem and the Intermediate Value Theorem . . 40
    7. Proof of the Effective Weierstrass Theorem 44

Chapter 1. Further Topics in Computable Analysis 50
    Introduction 50

    1. C^n Functions, $1 \leq n \leq \infty$ 51
    2. Analytic Functions 59
    3. The Effective Modulus Lemma and Some of Its Consequences . . 64
    4. Translation Invariant Operators 69

Part II. The Computability Theory of Banach Spaces 75

Chapter 2. Computability Structures on a Banach Space 77
    Introduction 77

    1. The Axioms for a Computability Structure 80
    2. The Classical Case: Computability in the Sense of Chapter 0 . . 82
    3. Intrinsic L^p-computability 83

4. Intrinsic l^p-computability 85
 5. The Effective Density Lemma and the Stability Lemma 85
 6. Two Counterexamples: Separability Versus Effective Separability
    and Computability on $L^\infty[0, 1]$ 88
 7. Ad Hoc Computability Structures 90

Chapter 3. The First Main Theorem and Its Applications 93
 Introduction 93

 1. Bounded Operators, Closed Unbounded Operators 96
 2. The First Main Theorem 101
 3. Simple Applications to Real Analysis 104
 4. Further Applications to Real Analysis 108
 5. Applications to Physical Theory 115

Part III. The Computability Theory of Eigenvalues and Eigenvectors . . . 121

Chapter 4. The Second Main Theorem, the Eigenvector Theorem, and
           Related Results 123
 Introduction 123

 1. Basic Notions for Unbounded Operators, Effectively Determined
    Operators 125
 2. The Second Main Theorem and Some of Its Corollaries 128
 3. Creation and Destruction of Eigenvalues 130
 4. A Non-normal Operator with a Noncomputable Eigenvalue 132
 5. The Eigenvector Theorem 133
 6. The Eigenvector Theorem, Completed 139
 7. Some Results for Banach Spaces 142

Chapter 5. Proof of the Second Main Theorem 149
 Introduction 149

 1. Review of the Spectral Theorem 151
 2. Preliminaries 158
 3. Heuristics 166
 4. The Algorithm 175
 5. Proof That the Algorithm Works 177
 6. Normal Operators 184
 7. Unbounded Self-Adjoint Operators 187
 8. Converses 188

Addendum: Open Problems 192
Bibliography 195
Subject Index 201

Major Interconnections

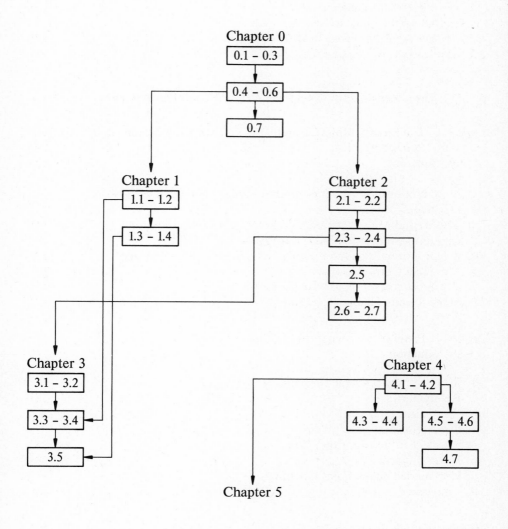

Introduction

This book deals with computable analysis. The subject, as its name suggests, represents a marriage between analysis and physics on the one hand, and computability on the other. Computability, of course, brings to mind computers, which are playing an ever larger role in analysis and physical theory. Thus it becomes useful to know, at least theoretically, which computations in analysis and physics are possible and which are not.

In this book, we attempt to develop a coherent framework for solving problems in this area. We will see that a variety of questions in computable analysis can be answered within this framework. For example, we will deal with computability for classical analysis, mathematical physics, Hilbert and Banach spaces, bounded and unbounded linear operators, eigenvalues and eigenvectors, and a variety of other topics. All of these are dealt with from the viewpoint of recursion theory, the theory of computability as treated in mathematical logic. Classical recursion theory provides a definition of computable function from integers to integers. Starting from this, the book develops corresponding notions of computability for real numbers, continuous functions, Hilbert space, L^p-spaces, and, more generally, arbitrary Banach spaces.

The framework used in this book is axiomatic. We axiomatize the notion of a "computability structure" on a Banach space. This allows a variety of applications to be treated under one heading. It is worth mentioning that the concept axiomatized is "computable sequence of vectors" of the Banach space. Then a point x is computable if the sequence $x, x, x, \ldots$ is computable. However, it is natural, and in fact necessary, to deal with sequences rather than individual points. For sequences lie at the center, both of recursion theory and analysis. A cornerstone of recursion theory is the notion of a recursive function, which is nothing more than a computable sequence of integers. In analysis, the topology on a Banach space is given by sequences.

We turn now to a discussion of some of the principal results in the book. These results are contained in Parts II and III. (We will discuss the contents of the more elementary Part I below.) There are three key results, the First and Second Main Theorems and the Eigenvector Theorem.

The First Theorem (in Chapter 3) asserts that, with certain mild side conditions, bounded operators preserve computability and unbounded operators do not. That is, a linear operator maps every computable element of its domain onto a computable element if and only if it is bounded. The side conditions are satisfied by all of the standard operators of analysis and physics. Hence we obtain a variety of applications merely by taking various well known linear operators and asking whether they are bounded or unbounded. Now, whether or not an operator is bounded, is a classical fact, which is usually well known. Thus, although the conclusion of the First Main Theorem is recursion-theoretic, the boundedness/unboundedness hypothesis is *not* dealt with in a recursion-theoretic manner. For this reason, the First Main Theorem is easy to apply.

Of course it is common practice for an analyst who is studying a particular linear operator to seek norms for which the operator is bounded. Our discussion indicates that this practice is not merely prudent, but necessary. In fact, the device of tailoring the norm to the problem at hand is seen—in a precise and not merely heuristic sense—as a necessary and sufficient means of preserving computability.

The Second Main Theorem (in Chapter 4) asserts that, under mild side conditions, a self-adjoint operator has computable eigenvalues, although the sequence of eigenvalues need not be computable. That is, although for each eigenvalue there is a program which computes it, different eigenvalues may require different programs. There is no master program which computes the entire sequence.

On the other hand, the Second Main Theorem does provide an explicit algorithm for computing individual eigenvalues. This algorithm applies to both bounded and unbounded self-adjoint operators. In particular, it applies to the standard operators of analysis and physics.

Incidentally, the side conditions needed for the Second Main Theorem are given in Chapter 4. Operators which satisfy these conditions are called *effectively determined*.

The Eigenvector Theorem (in Chapter 4), our third major result, asserts that there exists an effectively determined bounded self-adjoint operator T such that 0 is an eigenvalue of T of multiplicity one, but none of the eigenvectors corresponding to 0 is computable.

Relating the three major theorems, we note several contrasts. Begin with the First and Second Main Theorems. The first theorem asserts, in part, that unbounded operators do not preserve computability—i.e. that they map certain computable functions onto noncomputable functions. It may seem surprising, therefore, that all of the eigenvalues of an effectively determined unbounded self-adjoint operator are computable. In fact, the authors were surprised by this result. We had originally suspected that there should exist an effectively determined self-adjoint operator with noncomputable eigenvalues. In this connection, the fact that the eigenvalues *are* computable answers a question raised by Kreisel [1974].

There is a similar contrast between the Second Main Theorem and the Eigenvector Theorem, relating to the manner in which self-adjoint operators are used in quantum mechanics. As is well known, in quantum mechanics the operators correspond to "observables", the eigenvectors are associated with the states of the system, and the eigenvalues are related to the values actually measured. The Second Main Theorem asserts that the eigenvalues are computable, whereas the Eigenvector Theorem tells us that the eigenvectors need not be.

Introduction

It is time to spell out a little more fully the contents of the book. We begin with Chapter 2, which deals with the axioms, still postponing our discussion of the elementary Chapters 0 and 1. The axioms play an important role throughout the book. As noted above, the concept axiomatized is "computable sequence of vectors". The axioms relate to the principal structures on a Banach space. We recall that a Banach space is a linear space, which is endowed with a norm and which is complete in terms of this norm. The axioms relate the notion of computability to the three basic concepts of Banach space theory. Thus there are three axioms—one for linearity, one for limits, and one for the norm. When viewed in this light, the axioms appear to be minimal. It turns our that, in most of the interesting cases, the axioms are also maximal. That is, under mild side conditions, the axioms determine the computability structure uniquely. We emphasize that these axioms define a "computability structure" on a preexisting Banach space. We do *not* define a "computable Banach space". All of these matters are discussed fully in the introduction to Chapter 2.

In Chapter 3 we prove the First Main Theorem and give a variety of applications of it. These applications are of two types. In the first type, we apply the First Main Theorem directly to some linear operator. For example, by this means, we show that the heat and potential equations preserve computability, but that the wave equation does not. Similarly, we determine precisely the cases in which the Fourier transform preserves computability and the cases in which it does not. (In particular, the Fourier transform preserves computability from L^2 to L^2, giving an effective version of the Plancherel Theorem.) A host of other applications can be given, merely by taking various standard linear operators of analysis or physics and applying the First Main Theorem to them. The second type of application deals with problems which do not appear to involve linear operators at all. An example is the clasification of those step functions which are L^p-computable ($p < \infty$). Here, although the statement of the theorem does not involve linear operators, the proof does. In fact, in these cases, the introduction of a suitable linear operator—to which the First Main Theorem can be applied—provides the key to the proof.

Chapter 4 deals with topics surrounding the Second Main Theorem and the Eigenvector Theorem. These two theorems have been discussed at some length above. The Second Main Theorem has a number of corollaries. For example, under the hypotheses of that theorem, there exists a bounded operator whose norm is a noncomputable real. However, if in addition the operator is compact, then the norm is computable, and moreover the entire sequence of eigenvalues is computable. It could be asked whether the Second Main Theorem extends to operators which are not self-adjoint, or to Banach spaces other than Hilbert space. The answer is no. We show by an example that even on Hilbert space, when the operator is not self-adjoint, then noncomputable eigenvalues can occur. The chapter contains a number of other results. Several of these are related to the Eigenvector Theorem. In particular, the lemmas used in proving that theorem provide a variety of techniques for dealing with computability questions for Hilbert and Banach spaces.

We remark that, although the Second Main Theorem is stated in Chapter 4, its proof is postponed until Chapter 5.

So far we have mostly discussed Parts II and III, which are the main parts of the book. Now we turn to the introductory Part I. Chapter 0 develops the computability theory of all of the topics, except differentiation, which occur in a standard undergraduate course in real analysis. It provides the basic prerequisites for reading research papers in computable analysis—at least in those cases where the reasoning is classical (see below). Chapter 1 treats differentiation, analytic functions, and a variety of more advanced topics. Both of these chapters are based on the standard Grzegorczyk/Lacombe definition of a computable continuous function, and they make no use of Banach spaces or axioms. Nevertheless, as we will see in Chapter 2, the Grzegorczyk/Lacombe definition does fit naturally into the Banach space framework of Parts II and III.

The book contains a brief Addendum which discusses some open problems.

Reviewing the contents of these chapters as outlined above, we observe that most of the results are not simply effectivizations of classical theorems. Nor are they counterexamples to such effectivizations. In this sense, our results have no classical analog. Of course, there are exceptions. For example, the effective Plancherel theorem, mentioned above, is clearly the effectivization of a classical theorem. In the same vein, the noncomputability of solutions of the wave equation means that a certain existence theorem fails to effectivize. However, most of our results do not fit this format. For instance, as we saw above, the First Main Theorem involves a combination of classical and recursion theoretic hypotheses, although it leads to recursion theoretic conclusions. The same can be said of the Second Main Theorem, the Eigenvector Theorem, and, in fact, most of the results in the book.

We observe that the reasoning in this book is classical. Recall that, at the outset, we mentioned the general problem of deciding which computations in analysis and physics are possible and which are not. From this perspective, it is natural to reason classically, as analysts and physicists do. In particular, we do *not* work within the intuitionist or constructivist framework—e.g. the framework of Brouwer or Bishop. For, just as classical recursion theory allows the use of nonconstructive methods to study computability, our approach to recursive analysis does likewise. Our objective is to delineate the class of computable processes within the larger class of all processes. In this, our viewpoint is analogous to that of the complex analyst, who regards the analytic functions as a special case of the class of all functions, but regards all functions as existing mathematical objects. Of course, we do not wish to deny the value of constructive modes of reasoning. Our purpose here is simply to state, as clearly as possible, the viewpoint adopted in this book.

We have deliberately written the book so as to require a minimal list of prerequisites. As noted above, all of the work in the book is based on the standard notion of a recursive function from $\mathbb{N}$ to $\mathbb{N}$. From logic, we require only a few standard facts about recursive functions. These facts are spelled out in a section entitled *Prerequisites from Logic and Analysis*. All further recursion-theoretic notions, such as the Grzegorczyk/Lacombe definition of computability for continuous functions, are developed from scratch. From analysis we need the following. In Part I, we require only standard calculus. In Parts II and III, some familiarity with the L^p-spaces, Hilbert space, and Banach space is required. While the analysis which we use goes further than this, all additional analytic concepts are defined, frequently with discussion and examples. In the same vein, because the book is written for a mixed audience—including analysts and logicians—we have taken great pains to make our proofs clear and complete.

It may seen surprising to see an axiomatic approach used in connection with computability. Of course, axioms provide generality. The question is whether, by adopting an axiomatic approach, we lose the intrinsic quality traditionally associated with computability. In traditional recursion theory, for example, the notion of a recursive function from $\mathbb{N}$ to $\mathbb{N}$ is intrinsic. Although many different formulations have been given, they all lead to the same definition of computability. Actually, this intrinsic quality is largely preserved under the axiomatic approach. For, as mentioned above, under mild side conditions, the axioms determine the computability structure uniquely. Of course, the structure will be different for different Banach spaces. Yet, in each case, the axiomatic structure coincides with the natural one which has been extensively studied. For example, when applied to the Banach space of continuous functions on a closed interval, the axiomatic structure coincides with the classical Grzegorczyk/Lacombe definition of computability. For the L^p-spaces, p a computable real, $1 \leqslant p < \infty$, the axiomatic structure coincides with the natural definition of L^p-computability. (L^p-computability is defined in the obvious way—by taking the effective closure in L^p-norm of the Grzegorczyk/Lacombe computable functions.) Even in cases where relatively little work has been done up to now (e.g. Sobolev spaces) the axioms seem to provide a natural starting point. For more details, cf. Chapters 2 and 3.

Returning to L^p-computability, it turns out that certain discontinuous functions, and in particular certain step functions, are L^p-computable. At first glance, this seems to contradict a long standing perception that a computable function must be continuous. However, this perception depends implicitly on the assumption that the function is to be evaluated pointwise. In L^p theory—even classically—a function is never evaluated at a single point. For, as is well known, an L^p function is only defined up to sets of measure zero. So, instead of using pointwise evaluation, we use the L^p norm. Then the L^p-computability of certain step functions emerges, not as an ad hoc postulate, but as a consequence of the basic definition.

Although so far we have mainly discussed L^p spaces, many other computability structures on Banach spaces are related in a similar way to the standard Grzegorczyk-Lacombe definition. This holds, for example, for the energy norm and Sobolev spaces discussed in Chapter 3. However, there are other cases, e.g. those involving "ad hoc" computability structures, which bear no relationship to the Grzegorczyk-Lacombe definition. All of these cases, whether intrinsic or ad hoc, satisfy the axioms for a computability structure.

Although self-contained, the work in this monograph does not appear in a vacuum. There is a long tradition of research in recursive analysis using classical reasoning. Among those who have worked within this tradition are: Aberth, Grzegorczyk, Kreisel, Lachlan, Lacombe, Mazur, Metakides, Moschovakis, Mostowski, Myhill, Nerode, Rice, Robinson, Rogers, Shepherdson, Shore, Simpson, and Specker. Their work is cited in various places in the text, and also in the bibliography.

In the opinion of the authors, the field of computable analysis is in its infancy. There are numerous open problems, some hard and some easier. A small sample of these is given in the Addendum. Our hope is that this brief monograph will provide an easy introduction to the subject.

Prerequisites from Logic and Analysis

We begin with logic.

This book does not require any prior knowledge of formal logic.

First we expand upon some points made in the introduction to the book. All of our notions of computability (for real numbers, continuous functions, and beyond) are based on the notion of a recursive function $a: \mathbb{N} \to \mathbb{N}$ or $a: \mathbb{N}^q \to \mathbb{N}$. ($\mathbb{N}$ denotes the set of non-negative integers.) On the other hand, this book does not require a detailed knowledge of recursion theory. For reasons to be explained below, an intuitive understanding of that theory will suffice.

We continue for now to consider only functions from $\mathbb{N}$ to $\mathbb{N}$. Intuitively, a recursive function is simply a "computable" function. More precisely, a recursive function is a function which is computable by a Turing machine. The weight of fifty years experience leads to the conclusion that "recursive function" is the correct definition of the intuitive notion of a computable function. The definition is as solid as the definition of a group or a vector space. By now, the theory of recursive functions is highly developed.

However, as we have said, this book does not require a detailed knowledge of that theory. We avoid the need for heavy technical prerequisites in two ways.

1. Whenever we prove that some process is computable, we actually give the algorithm which produces the computation. As we shall see, some of these algorithms are quite intricate. But each of them is built up from scratch, so that the book is self-contained.
2. To prove that certain processes are not computable, we shall find that it suffices to know one basic result from recusive function theory—the existence of a recursively enumerable nonrecursive set. This we now discuss.

Imagine a computer which has been programmed to produce the values of a function a from nonnegative integers to nonnegative integers. We set the program in motion, and have the computer list the values $a(0), a(1), a(2), \ldots$ in order. This set A of values, a subset of the natural numbers, is an example of a "recursively enumerable set". If we take a general all purpose computer—e.g. a Turing machine —and consider the class of all such programs, we obtain the class of all recursively enumerable sets.

Suppose now that we have a recursively enumerable set A. Can we find an effective procedure which, for arbitrary n, determines whether or not $n \in A$? If $n \in A$, then clearly we have a procedure which tells us so. Namely, we simply list the values $a(0)$, $a(1)$, $a(2)$, ..., and if $n \in A$, then the value n will eventually appear. The difficulty comes if $n \notin A$. For then it seems possible that we might have no method of ascertaining this fact. We can, of course, list the set A to an arbitrarily large finite number of its elements, and we can observe that the value n has not occurred *yet*. However, we might have no way of determining whether n will show up at some later stage.

It turns out that, in general, there is no effective procedure which lists the elements $n \notin A$. Thus, there is no effective procedure which answers, for every n, the question: is $n \in A$?

On the other hand, for some sets A, the question of membership in A can be effectively answered. This is true for most of the sets of natural numbers commonly encountered in number theory, e.g. the set of primes. Such sets are called "recursive". By contrast, those sets A for which we can ascertain when $n \in A$ but not when $n \notin A$, are called "recursively enumerable nonrecursive". It is a fundamental result of logic that such sets exist (Proposition A below).

We now spell out these definitions and results in a formal manner. These are the only recursion-theoretic facts which are used in this book.

A set $A \subseteq \mathbb{N}$ is called *recursively enumerable* if $A = \phi$ or A is the range of a recursive function a. When A is infinite, the function a can be chosen to be one-to-one.

A set $A \subseteq \mathbb{N}$ is called *recursive* if both A and its complement $\mathbb{N} - A$ are recursively enumerable.

A cornerstone of recursion theory is the following result. For a proof see e.g. Kleene [1952], Rogers [1967], Davis [1958], Cutland [1980], Soare [1987].

Proposition A. *There exists a set $A \subseteq \mathbb{N}$ which is recursively enumerable but not recursive.*

Occasionally we will have to use a slightly stronger result—the existence of a recursively inseparable pair of sets.

Proposition B. *There exists a recursively inseparable pair of sets, i.e. a pair of subsets A, B of $\mathbb{N}$ such that*:
a) A and B are recursively enumerable.
b) $A \cap B = \phi$.
c) There is no recursive set C with $A \subseteq C$ and $B \subseteq \mathbb{N} - C$.

We need one further result from logic—the existence of a recursive pairing function J from $\mathbb{N} \times \mathbb{N} \to \mathbb{N}$, together with recursive inverse functions $K: \mathbb{N} \to \mathbb{N}$ and $L: \mathbb{N} \to \mathbb{N}$ such that:

$$J[K(n), L(n)] = n.$$

In fact, a standard form for J is:

$$J(x, y) = \frac{(x + y)(x + y + 1)}{2} + x$$

A technical comment. When we use the *characteristic function* of a set, we follow the custom of analysts rather than that of logicians. Thus we define the characteristic function χ_S of a set S by

$$\chi_S(x) = \begin{cases} 1 & \text{for } x \in S, \\ 0 & \text{for } x \notin S. \end{cases}$$

Now we turn to analysis. We assume the contents of a standard undergraduate course in real variables, together with the rudiments of measure theory. We also need the following well known definitions.

A *Banach space* is a real or complex vector space with a norm $\| \ \|$ such that:

$$\|x + y\| \leq \|x\| + \|y\|,$$

$$\|\alpha x\| = |\alpha| \cdot \|x\| \quad \text{for all scalars } \alpha,$$

$$\|x\| \geq 0 \quad \text{with equality if and only if } x = 0,$$

and such that the space is complete in the metric $\|x - y\|$.

A *Hilbert space* is a Banach space in which the norm is given by an inner product (x, y): the form (x, y) is linear in the first variable and conjugate linear in the second variable, and it is related to the norm by $\|x\| = (x, x)^{1/2}$.

The following spaces are used frequently in this book. Except for $C^\infty[a, b]$, they are all Banach spaces.

$L^p[a, b]$: for $1 \leq p < \infty$, the space of all measurable functions f on $[a, b]$ for which the L^p-norm $\|f\|_p = \left(\int_a^b |f(x)|^p \, dx \right)^{1/p}$ is finite.

l^p: for $1 \leq p < \infty$, the space of all real or complex sequences $\{c_n\}$ for which the l^p-norm $\|\{c_n\}\|_p = \left(\sum_{n=0}^\infty |c_n|^p \right)^{1/p}$ is finite.

$C[a, b]$: the space of all continuous functions f on $[a, b]$, endowed with the uniform norm $\|f\|_\infty = \sup_x \{|f(x)|\}$.

$C^n[a, b]$: the space of all n times continuously differentiable functions f on $[a, b]$; here the norm can be taken as the sum of the uniform norms of $f^{(k)}$, $0 \leq k \leq n$.

$C^\infty[a, b]$: the space of infinitely differentiable functions on $[a, b]$.

By obvious modifications, the interval $[a, b]$ can be replaced by the real line $\mathbb{R}$ or euclidean q-space $\mathbb{R}^q$. For the space $C(\ )$, we use $C_0(\mathbb{R}^q)$, the space of continuous functions which vanish at infinity.

Finally, in the important case where $p = 2$, L^p and l^p are Hilbert spaces.

Part I

Computability in Classical Analysis

Chapter 0
An Introduction to Computable Analysis

Introduction

This chapter is, in a sense, a primer of computable real analysis. We present here most of the basic methods needed to decide standard questions about the computability of real valued functions. Thus the chapter deals in a systematic way with the computability theory of real numbers, real sequences, continuous functions, uniform convergence, integration, maxima and minima, the intermediate value theorem, and several other topics. In fact, most of the usual topics from the standard undergraduate real variables course are treated within the context of computability.

There is one conspicuous omission. We postpone the treatment of derivatives until Chapter 1. This is because the computability theory of derivatives is a bit more complicated.

Section 1 deals with computable real numbers. A real number is computable if it is the effective limit of a computable sequence of rationals. All of these terms are defined at the beginning of the section. Surprisingly enough, many of the questions concerning individual computable reals require the consideration of computable sequences of reals. This topic is postponed until Section 2. However, a few questions can be answered at this preliminary stage. For example, the definition of "computable real" involves the notion of effective convergence for rational sequences. We show that the hypothesis of effective convergence is not redundant, by showing that there exist computable sequences of rationals which converge, but not effectively (Rice [1954], Specker [1949]). This is done by means of the Waiting Lemma, a standard recursion-theoretic result, which for the sake of completeness we prove. The Waiting Lemma is used repeatedly throughout the book. The section concludes with a brief discussion of some alternative definitions of "computable real number".

Section 2 deals with computable sequences of real numbers. This notion is essential for the entire book. To do work in computable analysis, one has to be totally fluent with all of the nuances related to computable sequences.

[The importance of sequences, rather than individual elements, reappears in Part II of this book. Thus in Chapter 2, where we lay down axioms for computability on a Banach space, the concept which is axiomatized is "computable sequence of vectors" in the Banach space.]

Section 2 begins with a careful discussion of the two notions: "computable sequence of reals" and "effective convergence of a double sequence $\{x_{nk}\}$ to a limit sequence $\{x_n\}$". The second of these has subtleties of its own. In particular, we must make the distinction between effective convergence (a notion from logic) and uniform convergence (a notion from analysis). In general neither implies the other. This is illustrated by Examples 1 and 2.

The section also contains two results, Propositions 1 and 2, dealing with effective convergence. As a corollary of these we obtain a noncomputable real which is the noneffective limit of a computable sequence of rationals (Rice [1954], Specker [1949]). The section concludes with two further examples, Examples 3 and 4, which serve several purposes. Firstly, these examples illustrate techniques which are echoed many times throughout the book. Secondly, the examples have corollaries which are of independent interest. For instance, there exists a sequence of rational numbers which is computable as a sequence of reals, but not as a sequence of rationals.

Section 3 gives the definition of a computable function of one or several real variables. In fact, it gives two equivalent definitions. Definition A is based on the pioneering work of Grzegorczyk [1955, 1957] and Lacombe [1955a, 1955b]. This definition involves a natural effectivization of basic constructs in analysis, and it is readily applicable to work in analysis. Definition B, due to Caldwell and Pour-El [1975], is based on an effectivization of the Weierstrass Approximation Theorem. For certain purposes—e.g. the treatment of integration over irregular domains—Definition B is more efficient than Definition A. Finally Definitions A and B, given for a single function defined on a bounded rectangle, are extended to sequences of functions and to unbounded domains (Definitions A', B', A", B"). It should be added that, over the years, a variety of definitions all equivalent to Definitions A and B have been given. These include definitions based on recursive functionals—cf. Grzegorczyk [1955, 1957] and Lacombe [1955a, 1955b] for details.

Here we reach a transition point. The basic definitions of this chapter have all been given, and now we begin to investigate systematically their consequences.

Section 4 deals with three elementary topics—composition of functions, patching, and extension of functions. The proofs in this section are worked out in great detail. Later on in the chapter we adopt a lighter style.

Section 5 treats two basic constructs in analysis—uniform convergence and integration. We first prove that the computable functions are closed under effective uniform convergence. The treatment of integration is, of necessity, more complicated. The reason is this. In order to handle the deeper problems concerned with integration, we need Definition B. However, in order to prove the equivalence of Definitions A and B, we need a preliminary result about integration (Theorem 5). From this we can deduce the equivalence of Definitions A and B (Theorem 6). Then follows a detailed discussion of the types of integrals which occur routinely in analysis. These include indefinite integrals, integrals depending on a parameter, a variety of line and surface integrals, and integrals over irregular domains. Such integrals appear at several places throughout this book. For instance, in later chapters we deal with the Cauchy integral formula, Kirchhoff's solution formula

1. Computable Real Numbers

for the wave equation, the corresponding formula for the heat equation, and others. A general result which encompasses all integrals of this type is given in Corollary 6c.

[Part of the proof of Theorem 6 is deferred until Section 7—see below. Of course, this proof uses only results proved prior to the statement of Theorem 6.]

Section 6 deals with the max-min theorem, the intermediate value theorem, and certain other topics. Both the max-min theorem and the intermediate value theorem effectivize for a single computable function. However, when we consider computable sequences of functions we see a divergence: the max-min theorem effectivizes for sequences, whereas the intermediate value theorem does not. There is a further subtlety associated with the max-min theorem. Although the maximum value taken by a computable function is computable, the point where this maximum occurs need not be (Kreisel [1958], Lacombe [1957b], Specker [1959]). We omit the proof of this theorem, since we do not need it. Two further topics are treated. As a corollary of the intermediate value theorem, we prove that the computable reals form a real closed field. The section concludes with a brief discussion of the Mean Value Theorem.

Section 7 completes the proof of Theorem 6 (equivalence of Definitions A and B). This is done by giving an effective version of the Weierstrass Approximation Theorem. The proof is rather complicated, and is included partly as an illustration of technique. A much easier proof is given in Chapter 2, Section 5.

1. *Computable Real Numbers*

While every rational number is computable, it is clear that not every real number is. For the set of all computer programs is countable, whereas the set of real numbers is not. Roughly speaking, a computable real is one which can be effectively approximated to any desired degree of precision by a computer program given in advance. Thus the number π is computable, since there exist finite recipes for computing it. When more precision is desired the computation may take longer, but the recipe itself does not change.

In this section, we define "computable real" and prove some simple results about effective and noneffective convergence. As remarked in the introduction to the chapter, further results on individual computable reals require a knowledge of "computable sequences of reals", and hence are postponed until Section 2.

We begin with the fact that a real number is the limit of a Cauchy sequence of rationals. There are two aspects to the effectivization of this concept: 1) the sequence of rationals must be computable, and 2) the convergence of this sequence to its limit must be effective. We now examine each of these requirements in turn.

For 1) we mean—as already suggested above—that the entire sequence of rationals is computed by a finite set of instructions given in advance. For 2) we mean that there is a second set of instructions, also given in advance, which will tell us, for any $\varepsilon > 0$, a point where an error less than ε has been achieved. The precise definitions are:

Definition 1. A sequence $\{r_k\}$ of rational numbers is *computable* if there exist three recursive functions a, b, s from $\mathbb{N}$ to $\mathbb{N}$ such that $b(k) \neq 0$ for all k and

$$r_k = (-1)^{s(k)} \frac{a(k)}{b(k)} \quad \text{for all } k.$$

Definition 2. A sequence $\{r_k\}$ of rational numbers *converges effectively* to a real number x if there exists a recursive function $e: \mathbb{N} \to \mathbb{N}$ such that for all N:

$$k \geq e(N) \quad \text{implies} \quad |r_k - x| \leq 2^{-N}.$$

Definition 3. A real number x is *computable* if there exists a computable sequence $\{r_k\}$ of rationals which converges effectively to x.

A complex number is called computable if its real and imaginary parts are computable. Similarly, a q-vector $(x_1, \ldots, x_q)$ is computable if each of its components is computable.

Obviously, the notion of a computable real is central to recursive analysis. Eventually, in order to obtain far-reaching results, we shall have to generalize this notion to sequences of reals, continuous functions, and beyond. Nevertheless, certain basic questions already appear at this stage.

Proposition 0. *Let x be a computable real number. If $x > 0$, then there is an effective procedure which shows this. Likewise for $x < 0$. If $x = 0$, there is in general no effective way of proving this.*

Proof. Since x is a computable real, x is the limit of a computable sequence $\{r_k\}$ of rationals, with an effective modulus of convergence $e(N)$ as above. Suppose that $x > 0$. Then the following procedure will eventually terminate and provide an effective proof that $x > 0$:

For $N = 0, 1, 2, \ldots$, compute $e(N)$ and $r_{e(N)}$, and wait until an $r_{e(N)}$ turns up with

$$r_{e(N)} > 2^{-N}.$$

If $x > 0$, such an N must eventually occur. For suppose $2^{-N} < x/2$. Then, since $|r_{e(N)} - x| \leq 2^{-N} < x/2$, $r_{e(N)} > x/2 > 2^{-N}$.

In the reverse direction, since $|r_{e(N)} - x| \leq 2^{-N}$, the condition $r_{e(N)} > 2^{-N}$ implies $x > 0$.

Similarly for $x < 0$.

However, such a simple test will not work for the case where $x = 0$. In fact, there is no effective test for this case. A counterexample demonstrating this is given in the next section (Fact 3 following Example 4). □

Remark. The processes of analysis frequently require that comparisons be made between real numbers. Yet, as we have just seen, such comparisons cannot always

1. Computable Real Numbers

be made effectively. Often we can obtain an effective substitute by using rational approximations. Thus the following simple fact will be used constantly.

Consider a computable sequence $\{r_n\}$ of rationals: $r_n = (-1)^{s(n)}[a(n)/b(n)]$ as in Definition 1. Then $r_n = 0$ if and only if $a(n) = 0$; $r_n > 0$ if and only if $a(n) > 0$ and $s(n)$ is even; and $r_n < 0$ if and only if $a(n) > 0$ and $s(n)$ is odd. This gives an effective test for picking out the sequences $\{n: r_n = 0\}$, $\{n: r_n > 0\}$, and $\{n: r_n < 0\}$.

In a word: exact comparisons are possible for computable sequences of rationals, although not necessarily for computable reals.

We turn now to the question of effective convergence. Generally, when a computable sequence of rationals $\{r_k\}$ converges to a real number x, the convergence may or may not be effective. There is one important case where effectiveness of convergence is guaranteed; it corresponds to the nested intervals approach for the definition of a real number.

Let $\{a_k\}$ and $\{b_k\}$ be computable sequences of rationals which are monotone upwards and downwards respectively and converge to x: i.e. $a_0 \leqslant a_1 \leqslant \cdots$, $b_0 \geqslant b_1 \geqslant \cdots$, and $a_k \leqslant x \leqslant b_k$ for all k. Then these sequences converge effectively to x, as we now show.

The differences $(b_k - a_k)$ decrease monotonically to zero. Hence, to define an effective modulus of convergence $e(N)$, we simply wait, for each N, until an index $e(N)$ with $(b_{e(N)} - a_{e(N)}) \leqslant 2^{-N}$ turns up.

In the above situation, we had two monotone sequences, converging upwards and downwards respectively to the limit x. Suppose we merely have one computable sequence $\{s_k\}$ which converges upwards monotonically to x. Then the convergence need not be effective.

To show this—our first example of noneffective convergence—we need some preliminary results. These results will be used several times throughout this book.

Lemma (Waiting Lemma). *Let $a: \mathbb{N} \to \mathbb{N}$ be a one to one recursive function generating a recursively enumerable nonrecursive set A. Let $w(n)$ denote the "waiting time"*

$$w(n) = \max\{m: a(m) \leqslant n\}.$$

Then there is no recursive function c such that $w(n) \leqslant c(n)$ for all n.

Proof. The term "waiting time" contains the essential idea of the proof. If we could effectively bound the waiting time, then we would have a decision procedure for telling whether or not $n \in A$. Namely, wait until the waiting time for n has elapsed. If the value n has not turned up by this time then it never will. Now more formally:

Suppose, on the contrary, that $w(n) \leqslant c(n)$ with c recursive. Then the set A is recursive. Here is a decision procedure for A.

For any n, compute $a(m)$ for all $m \leqslant c(n)$. If one of these values $a(m) = n$, then $n \in A$; otherwise $n \notin A$.

To see this, we reason as follows. Obviously if $a(m) = n$ for some $m \leqslant c(n)$, then $n \in A$. Otherwise, since $c(n) \geqslant w(n)$, we have $a(m) \neq n$ for all $m \leqslant w(n)$. But $w(n)$ gives the last value of m for which $a(m) \leqslant n$, and hence the last value (if any) for which

$a(m) = n$. Thus if n has not turned up as a value of $a(m)$ by this time, then it never will. □

There is a close connection between the waiting time defined above and the modulus of convergence for certain series.

Lemma (Optimal Modulus of Convergence). *Let $a: \mathbb{N} \to \mathbb{N}$ be a one to one recursive function generating a recursively enumerable nonrecursive set A, and let w be the waiting time, as in the preceding lemma. Consider the series:*

$$s_k = \sum_{m=0}^{k} 2^{-a(m)},$$

and let $x = \lim_{k \to \infty} s_k$. Define the "optimal modulus of convergence" $e^*(N)$ to be the smallest integer such that

$$k \geqslant e^*(N) \quad \text{implies} \quad (x - s_k) \leqslant 2^{-N}.$$

Then $w(N) = e^*(N)$.

Proof. For any k,

$$x - s_k = \sum_{m=k+1}^{\infty} 2^{-a(m)}.$$

We will show that the waiting time $w(N)$ satisfies the conditions defining $e^*(N)$.

By definition of $w(N)$ as the last value m with $a(m) \leqslant N$, we have in particular that $a[w(N)] \leqslant N$.

Suppose $k < w(N)$. Then the series for $x - s_k$ contains the term with $m = w(N)$, and the value of this term is $2^{-a[w(N)]} \geqslant 2^{-N}$. Since the series also contains other positive terms, $x - s_k > 2^{-N}$.

Suppose $k \geqslant w(N)$. Then the series for $x - s_k$ (which begins with $m = k + 1$) contains no term $2^{-a(m)}$ with $a(m) \leqslant N$. Hence the series is dominated by $\sum_{a=N+1}^{\infty} 2^{-a} = 2^{-N}$, and $x - s_k \leqslant 2^{-N}$.

Hence $w(N)$ fulfills precisely the condition by which we defined $e^*(N)$. □

By combining the two previous lemmas, we have an example of a computable monotone sequence of rationals which converges, but does not converge effectively.

Example (Noneffective convergence, cf. Rice [1954], Specker [1949]). Let $a: \mathbb{N} \to \mathbb{N}$ be as in the two preceding lemmas, and let $\{s_k\}$ be the computable sequence given by:

$$s_k = \sum_{m=0}^{k} 2^{-a(m)}.$$

Then $\{s_k\}$ converges noneffectively to its limit x.

Proof. Since the function a is one to one, obviously the series converges. If the convergence were effective, there would be a recursive function $e(N)$ such that

$$k \geqslant e(N) \quad \text{implies} \quad (x - s_k) \leqslant 2^{-N}.$$

Comparing this with the "optimal" modulus of convergence in the previous lemma, we see that $e(N) \geqslant e^*(N)$.

By the previous lemma $e^*(N) = w(N)$, so that the recursive function $e(N) \geqslant w(N)$. By the Waiting Lemma, this is impossible. □

Note. Although the sequence s_k is computable, we cannot say that the limit x is computable, since the convergence is noneffective. However, x might still be computable, as there could be another computable sequence $\{r_k\}$ which converges effectively to x. In fact, as we shall show in the next section (Corollary 2b), this cannot happen: x is not computable.

Alternative definitions of "computable real number". All of the standard definitions of real number (in the classical, noncomputable sense) effectivize to give the same definition of "computable real". Specifically, we list the following four well known methods for constructing the reals from the rationals:

1) Cauchy sequences (the method effectivized in this section).
2) Dedekind cuts.
3) Nested intervals.
4) Decimals to the base b, an integer > 1.

The fact that the effective versions of 1–4 are equivalent was first proved by R.M. Robinson [1951]. Robinson observed that the key step is to show that, for any computable real α, the function $a(n) = [n\alpha]$ (where [ ] = greatest integer) is recursive. The equivalence of 1–4 follows easily from this observation. In this regard we cite another important early paper of Rice [1954]. See also Mazur [1963].

In this book we shall work exclusively with definition 1 (Cauchy sequences). The equivalence of the other definitions is not needed. For that reason we do not spell out any further details.

2. Computable Sequences of Real Numbers

A sequence of real numbers may not be computable, even though each of its individual elements is. With a finite sequence, of course, there is no problem, since finitely many programs can be combined into one. But for an infinite sequence $\{x_n\}$, we might have a program for each n, but no way to combine these infinitely many programs into one.

To say that a sequence is computable means that there is a master program which, upon the input of the number n, computes x_n.

The idea of sequential computability plays a key role in this book. Consequently, in this section, we give a thorough treatment of the points at issue. First, we lay down the basic definitions. These definitions involve the notion of effective convergence of a double sequence $\{x_{nk}\}$ to a sequence $\{x_n\}$. We include a discussion of the similarities and contrasts between this and the analyst's notion of "uniform convergence". Two examples (Examples 1 and 2) illustrate these points. Then follow two propositions—Proposition 1 (closure under effective convergence) and Proposition 2 (dealing with monotone convergence)—which are used so often in this book that eventually we stop referring to them by name and simply regard them as "well known facts". Proposition 2, incidentally, leads us for the first time to noncomputable reals (Corollary 2b). The section closes with two more counter-examples (Examples 3 and 4) and several "facts" which follow from them. We suggest that the techniques used in constructing the examples may be of interest in their own right. These techniques, refined and extended, recur at many points throughout the book.

We turn now to the basic definitions.

As with the case of computable reals in Section 1, there are two aspects to the definition of a "computable sequence of reals" $\{x_n\}$: 1) we require a computable double sequence of rationals $\{r_{nk}\}$ which converges to $\{x_n\}$ as $k \to \infty$, and 2) this convergence must be "effective". We now make these notions precise.

First a technicality. A double sequence will be called *computable* if it is mapped onto a computable sequence by one of the standard recursive pairing functions from $\mathbb{N} \times \mathbb{N}$ onto $\mathbb{N}$. Similarly for triple or q-fold sequences.

Definition 4. Let $\{x_{nk}\}$ be a double sequence of reals and $\{x_n\}$ a sequence of reals such that, as $k \to \infty$, $x_{nk} \to x_n$ for each n. We say that $x_{nk} \to x_n$ *effectively in k and n* if there is a recursive function $e: \mathbb{N} \times \mathbb{N} \to \mathbb{N}$ such that for all n, N:

$$k \geq e(n, N) \quad \text{implies} \quad |x_{nk} - x_n| \leq 2^{-N}.$$

(In words, the variable N corresponds to the error 2^{-N}, and the function $e(n, N)$ gives a bound on k sufficient to attain this error. Without loss of generality, we can assume that $e(n, N)$ is an increasing function of both variables.)

Combining Definition 4 with Definition 1 of the previous section, we obtain:

Definition 5. A sequence of real numbers $\{x_n\}$ is *computable* (as a sequence) if there is a computable double sequence of rationals $\{r_{nk}\}$ such that $r_{nk} \to x_n$ as $k \to \infty$, effectively in k and n.

The following variant is often useful.

Definition 5a. A sequence of real numbers $\{x_n\}$ is *computable* (as a sequence) if there is a computable double sequence of rationals $\{r_{nk}\}$ such that

$$|r_{nk} - x_n| \leq 2^{-k} \quad \text{for all } k \text{ and } n.$$

2. Computable Sequences of Real Numbers

Obviously, the condition in Definition 5a implies that in Definition 5. For the converse, we reason as follows. From Definitions 4 and 5, there is a computable sequence of rationals $\{r_{nk}\}$, and a recursive function $e(n, N)$, such that $k \geq e(n, N)$ implies $|r_{nk} - x_n| \leq 2^{-N}$. Then we simply replace $\{r_{nk}\}$ by the computable subsequence,

$$r'_{nk} = r_{n, e(n, k)},$$

to obtain $|r'_{nk} - x_n| \leq 2^{-k}$, as desired.

The above definitions extend in the obvious way to complex numbers and to q-vectors. Thus a sequence of complex numbers is called *computable* if its real and imaginary parts are computable sequences. A sequence of q-vectors is called *computable* if each of its components is a computable sequence of real or complex numbers.

There are subleties associated with the idea of effective convergence which will come up repeatedly throughout this work. Many of these will be dealt with as they appear. However, the following discussion clarifies one simple point.

There is a possible confusion, which comes to the forefront in this book, between the notion of "uniformity" as used in logic and in analysis. When an analyst says "$x_{nk} \to x_n$ as $k \to \infty$, uniformly in n" he or she means that the rate of convergence can be made *independent of n*. In logic, the same phrase would often mean *dependent on n, but in a computable way*. To avoid confusion, in this book we shall set the following conventions:

"uniformly in n" means independent of n;
"effectively in n" means governed by a recursive function of n.

Thus we use the word "uniformly" in the sense of analysis, and describe logical uniformities by the term "effective".

The following examples illustrate these distinctions.

Example 1. The double sequence

$$x_{nk} = \frac{k}{k + n + 1}$$

converges as $k \to \infty$ to the sequence $\{x_n\} = \{1, 1, 1, \ldots\}$. The convergence is not uniform in n. However, the convergence is effective in both k and n: an effective modulus of convergence (for the error 2^{-N}) is given by $e(n, N) = (n + 1) \cdot 2^N$.

Example 2. In the previous section, we gave an example of a sequence $\{s_k\}$ which converges noneffectively to its limit x. Suppose we set

$$x_{nk} = s_k \quad \text{for all } n.$$

Then $\{x_{nk}\}$ converges uniformly in n (since it does not depend on n), but noneffectively in k.

It is more difficult to give an example of a double sequence $\{x_{nk}\}$ which converges as $k \to \infty$, effectively in k but noneffectively in n. We shall give such an example later in this section.

We now give two propositions concerning computable sequences. These will be used repeatedly throughout the book.

Proposition 1 (Closure under effective convergence). *Let $\{x_{nk}\}$ be a computable double sequence of real numbers which converges as $k \to \infty$ to a sequence $\{x_n\}$, effectively in k and n. Then $\{x_n\}$ is computable.*

Proof. Since $\{x_{nk}\}$ is computable, we have by Definition 5a a computable triple sequence of rationals $\{r_{nkN}\}$ such that

$$|r_{nkN} - x_{nk}| \leq 2^{-N} \quad \text{for all } n, k, N.$$

Since $x_{nk} \to x_n$ effectively in k and n, there is a recursive function $e(n, N)$ such that

$$k \geq e(n, N) \quad \text{implies} \quad |x_{nk} - x_n| \leq 2^{-N}.$$

Then the computable double sequence of rationals,

$$r'_{nN} = r_{n, e(n, N), N},$$

satisfies

$$|r'_{nN} - x_n| \leq 2 \cdot 2^{-N},$$

whence $\{x_n\}$ is computable, as desired. □

Proposition 2 (Monotone convergence). *Let $\{x_{nk}\}$ be a computable double sequence of real numbers which converges monotonically upwards to a sequence $\{x_n\}$ as $k \to \infty$: i.e. $x_{n0} \leq x_{n1} \leq x_{n2} \leq \cdots$ and $x_{nk} \to x_n$ for each n. Then $\{x_n\}$ is computable if and only if the convergence is effective in both k and n.*

Corollary 2a. *If a computable sequence $\{x_k\}$ converges monotonically upwards to a limit x, then the number x is computable if and only if the convergence is effective.*

Corollary 2b. *There exists a computable sequence of rationals which converges to a noncomputable real.*

[In particular, this shows that the effective analog of the Bolzano-Weierstrass Theorem dos not hold.]

Proof of corollaries (assuming the proposition). Corollary 2a follows immediately by holding n fixed.

For Corollary 2b. In Section 1 we gave an example of a computable sequence $\{s_k\}$ of rational numbers which converges monotonically upwards to a limit x, but

2. Computable Sequences of Real Numbers

for which the convergence is not effective. Hence by Corollary 2a, the limit x is not computable. □

Proof of Proposition 2. The "if" part follows (without any assumption of monotonicity) from Proposition 1 above.

It is for the "only if" part that we need monotonicity. Suppose $\{x_n\}$ is computable. Since $\{x_{nk}\}$ is also computable, there exists a computable double sequence of rationals $\{r_{nN}\}$ and a computable triple sequence of rationals $\{r_{nkN}\}$ such that:

$$|r_{nN} - x_n| \leq 2^{-N}/6 \quad \text{for all } n, N;$$
$$|r_{nkN} - x_{nk}| \leq 2^{-N}/6 \quad \text{for all } n, k, N.$$

Now define the recursive function $e(n, N)$ to be the first index k such that

$$|r_{nkN} - r_{nN}| \leq 2^{-N}/2.$$

Such a k must exist, since $x_{nk} \to x_n$, and the sum of the two errors above is $2^{-N}/3$. Furthermore,

$$k = e(n, N) \quad \text{implies} \quad |x_{nk} - x_n| \leq \tfrac{5}{6} \cdot 2^{-N}.$$

Now the fact that this holds for all $k \geq e(n, N)$, and not merely for $k = e(n, N)$, follows from the monotonicity of $\{x_{nk}\}$ as a function of k. Thus $e(n, N)$ provides an effective modulus of convergence for the limit process $x_{nk} \to x_n$. □

The following remark is included for the sake of completeness.

Remark (Elementary functions). Let $\{x_n\}$ and $\{y_n\}$ be computable sequences of real numbers. Then the following sequences are computable:

$x_n \pm y_n$,

$x_n y_n$,

x_n/y_n ($y_n \neq 0$ for all n),

$\max(x_n, y_n)$ and $\min(x_n, y_n)$,

$\exp x_n$,

$\sin x_n$ and $\cos x_n$,

$\log x_n$ ($x_n > 0$ for all n),

$\sqrt[m]{x_n}$ ($x_n \geq 0$ for all n),

$\arcsin x_n$ and $\arccos x_n$ ($|x_n| \leq 1$ for all n),

$\arctan x_n$.

Thus, in particular, the computable reals form a field. We will show in Section 6 that they also form a real closed field.

Proof. Algorithms for doing these computations are so well known that we need not set them down here. We give one example—not even the most complicated one—because it illustrates the use of the "effective convergence proposition", Proposition 1 above.

First, the proofs of computability for addition, subtraction, and multiplication are routine. Then to prove the computability of $\{\exp x_n\}$, we use Taylor series. Let

$$s_{nk} = \sum_{j=0}^{k} (x_n^j/j!).$$

The double sequence $\{s_{nk}\}$ is computable and converges to $\{\exp x_n\}$ as $k \to \infty$, effectively in k and n. Hence by Proposition 1, $\{\exp x_n\}$ is computable. □

The functions $\sin x$, $\cos x$, arc $\sin x$, and arc $\cos x$ can all be handled similarly, by Taylor series which converge effectively over the entire domain of definition of the function. For arc tan x, we use the identity arc tan $x = $ arc sin $[x/(1 + x^2)^{1/2}]$. Now $\log x$ and $\sqrt[m]{x}$ require a little more work, since their Taylor series have limited domains of convergence. However, the detailed treatment of these functions is mundane, and we shall not spell it out.

Incidentally, the computability of arc tan x_n implies the computability of π.

The last two examples in this section serve two purposes. First, they introduce techniques which will be used repeatedly throughout the book. Second, they have the merit of answering, at the same time, four different questions concerning the topics in this section. These results will appear as "facts" at the end.

In each of these examples, $a: \mathbb{N} \to \mathbb{N}$ is a one to one recursive function generating a recursively enumerable nonrecursive set A.

Example 3. Consider the computable double sequence $\{x_{nk}\}$ defined by:

$$x_{nk} = \begin{cases} 1 & \text{if } n = a(m) \text{ for some } m \leq k, \\ 0 & \text{otherwise.} \end{cases}$$

Then as $k \to \infty$, $x_{nk} \to x_n$ where:

$$x_n = 1 \quad \text{if } n \in A, \qquad 0 \quad \text{if } n \notin A.$$

Thus $\{x_n\}$ is the characteristic function of the set A.

Now $\{x_n\}$ is not a computable sequence of reals. For it if were, then by approximating the x_n effectively to within an error of $1/3$, we would have an effective procedure to determine the integers $n \in A$ and also $n \notin A$. Thus the set A would be recursive, a contradiction.

2. Computable Sequences of Real Numbers

We mention in passing that, although $\{x_n\}$ is not a computable sequence of real numbers, its individual elements are computable reals—in fact, they are either 0 or 1.

Our last example is a modification of the preceding one,

Example 4. Consider the computable double sequence $\{x_{nk}\}$ defined by:

$$x_{nk} = \begin{cases} 2^{-m} & \text{if } n = a(m) \text{ for some } m \leq k, \\ 0 & \text{otherwise.} \end{cases}$$

Then as $k \to \infty$, $x_{nk} \to x_n$ where:

$$x_n = \begin{cases} 2^{-m} & \text{if } n = a(m) \text{ for some } m, \\ 0 & \text{if } n \notin A. \end{cases}$$

The case description of $\{x_n\}$ above is not effective, since it requires knowledge of $a(m)$ for infinitely many m. Nevertheless, $\{x_n\}$ is a computable sequence of reals. For the convergence of x_{nk} to x_n is effective in k and n. To see this, we observe that the only case where $x_{nk} \neq x_n$ occurs when $n = a(m)$ for some $m > k$. Then $x_{nk} = 0$, and $x_n = 2^{-m} < 2^{-k}$. Hence:

$$|x_{nk} - x_n| < 2^{-k} \qquad \text{for all } k, n.$$

Hence by Proposition 1, $\{x_n\}$ is computable.

The following facts are consequences of Examples 3 and 4 above.

Fact 1. Let A be a recursively enumerable non recursive set, and let $\chi(n)$ be the characteristic function of A. Then there exists a computable double sequence $\{x_{nk}\}$ which converges (noneffectively) to $\chi(n)$ as $k \to \infty$.

This follows from Example 3, with $\chi(n)$ in place of x_n.

In Example 2 we gave an instance of a computable double sequence $\{x_{nk}\}$ which converges to a sequence $\{x_n\}$ as $k \to \infty$, effectively in n but noneffectively in k. Now we reverse the situation:

Fact 2. There exists a computable double sequence $\{x_{nk}\}$ which converges to a sequence $\{x_n\}$ as $k \to \infty$, effectively in k but noneffectively in n.

Again this follows from Example 3. Consider the $\{x_{nk}\}$ and $\{x_n\}$ given there. Fix n. Then the convergence of $x_{nk} \to x_n$ as $k \to \infty$ is effective in k; indeed, $x_{nk} = x_n$ for all but finitely many k. But the convergence is not effective in n. For if it were, then by Proposition 1 above, the limit sequence $\{x_n\}$ would be computable.

The next result substantiates an assertion made in Proposition 0 of Section 1.

Fact 3. The condition $x = 0$ for computable real numbers cannot be decided effectively.

Here we use Example 4. The sequence $\{x_n\}$ given there is a computable sequence of real numbers. Yet the set $\{n: x_n = 0\}$, which equals the set $\{n: n \notin A\}$, cannot be effectively listed.

Fact 4. There exists a sequence $\{x_n\}$ of rational numbers which is computable as a sequence of reals, but not computable as a sequence of rationals.

Again the sequence $\{x_n\}$ from Example 4 suffices. We have seen that $\{x_n\}$ is a computable sequence of reals. But $\{x_n\}$ is not a computable sequence of rationals— i.e. x_n cannot be expressed in the form $x_n = (-1)^{s(n)}[p(n)/q(n)]$ with recursive functions p, q, s. For if it could, then the condition $x_n = 0$ would be effectively decidable. As we saw in the discussion following Fact 3, this is not so.

Alternative definitions of "computable sequence of real numbers". This discussion parallels a corresponding discussion at the end of Section 1. However, there are striking differences in the results, as we shall see. In Section 1 we noted that, for a single computable real, there are numerous equivalent definitions. Thus the definitions via 1. Cauchy sequences (as in this book), 2. Dedekind cuts, 3. nested intervals, and 4. decimals to the base b, all effectivize to give equivalent definitions of "computable real number". However, Mostowski [1957] showed by counterexamples that the corresponding definitions for *sequences* of real numbers are *not* equivalent. He observed that the Cauchy definition is presumably the correct one.

There are several reasons for preferring the Cauchy definition. We mention the following in passing. Suppose we took any of the *other* above-mentioned definitions (e.g. via Dedekind cuts) for "computable sequence of reals". Then we would obtain some rather bizarre results, such as: There exist "computable" sequences $\{x_n\}$ and $\{y_n\}$ whose sum $\{x_n + y_n\}$ is not "computable". Of course, this does not happen with the Cauchy definition.

[The above-mentioned counterexample will be used nowhere in this book. Nevertheless, we give a brief explanation of it. Consider e.g. the Dedekind definition. It is possible to give an example of a Cauchy computable sequence $\{z_n\}$ which is not Dedekind computable, and in which the elements z_n are rational numbers. On the other hand, it can be shown that if $\{x_n\}$ is Cauchy computable, and the values x_n are irrational, then $\{x_n\}$ is Dedekind computable. Now, starting with the above example $\{z_n\}$, set $x_n = z_n + \sqrt{2}$, $y_n = -\sqrt{2}$. Then $\{x_n\}$ and $\{y_n\}$ are Cauchy computable, and they take irrational values. Hence $\{x_n\}$ and $\{y_n\}$ are Dedekind computable, but $\{x_n + y_n\} = \{z_n\}$ is not.]

3. Computable Functions of One or Several Real Variables

We begin with a bit of historical background. In recursion-theoretic practice, a real number x is usually viewed as a function $a: \mathbb{N} \to \mathbb{N}$. Then a function of a real variable $f(x)$ is viewed as a functional, i.e. a mapping Φ from functions a as above into similar functions b. Within this theory, there is a standard and well explored notion of a

3. Computable Functions of One or Several Real Variables

recursive functional, investigated by Kleene and others. We shall not define these terms here, since we will not need them. Suffice it to say that these notions gave the original definition of "computable function of a real variable": the real-valued function f is called *computable* if the corresponding functional Φ is recursive (Lacombe [1955a, 1955b], Grzegorczyk [1955, 1957]).

From a recursion-theoretic viewpoint, this definition appears to capture the notion of "computability" for real-valued functions very well. However, this approach is not readily amenable to work in analysis. For an analyst does not view a real number as a function $a: \mathbb{N} \to \mathbb{N}$, nor a function of a real variable as a functional from such functions a into other functions b. Accordingly, much of the intuition of the analyst is lost. Clearly it is desirable to have a definition of "computable function of a real variable" equivalent to the recursion-theoretic one, but couched in the traditional notions of analysis.

Such a definition was provided by Grzegorczyk [1955, 1957]. It is equivalent, as Grzegorczyk proved, to the recursion-theoretic definition, but expressed in analytic terms. This is definition A below.

From the point of view of the analyst, Definition A is quite natural. For a real function f is determined if we know (a) the values of f on a dense set of points, and (b) that f is continuous. Definition A simply effectivizes these notions. Condition (i) in Definition A effectivizes (a), and condition (ii) effectivizes (b).

In this book, we give another equivalent definition based on an effective version of the Weierstrass Approximation Theorem (Definition B). This definition is useful in many applications. The equivalence of Definitions A and B is proved in Sections 5 and 7.

One final note. The equivalent Definitions A and B below are the natural ones for continuous functions. However, there are other definitions which apply to more general classes of functions—e.g. L^p functions. In fact, they apply to arbitrary Banach spaces. These definitions will be introduced in Chapter 2. They will play a major role in later chapters of the book. We will see, however, that when the general definitions of Chapter 2 are applied to the special case of continuous functions, they reduce to Definitions A and B.

We turn now to the definitions themselves.

For simplicity, we first consider the case where the function f is defined on a closed bounded rectangle I^q in $\mathbb{R}^q$. Specifically, $I^q = \{a_i \leq x_i \leq b_i, 1 \leq i \leq q\}$, where the endpoints a_i, b_i are computable reals.

As noted above, the following definition is due to Grzegorczyk/Lacombe.

Definition A (Effective evaluation). Let $I^q \subseteq \mathbb{R}^q$ be a computable rectangle, as described above. A function $f: I^q \to \mathbb{R}$ is *computable* if:

(i) f is *sequentially computable*, i.e. f maps every computable sequence of points $x_k \in I^q$ into a computable sequence $\{f(x_k)\}$ of real numbers;

(ii) f is *effectively uniformly continuous*, i.e. there is a recursive function $d: \mathbb{N} \to \mathbb{N}$ such that for all $x, y \in I^q$ and all N:

$$|x - y| \leq 1/d(N) \quad \text{implies} \quad |f(x) - f(y)| \leq 2^{-N},$$

where | | denotes the euclidean norm.

Our second equivalent definition involves the notion of a computable sequence of polynomials. In the real case, which we are here considering, these polynomials can have rational coefficients.

By a *computable* sequence of rational polynomials, we mean a sequence

$$p_n(x) = \sum_{i=0}^{D(n)} r_{ni} x^i,$$

where $D: \mathbb{N} \to \mathbb{N}$ is a recursive function, and $\{r_{ni}\}$ is a computable double sequence of rationals.

The following definition is due to Caldwell and Pour-El [1975].

Definition B (Effective Weierstrass). Let $I^q \subseteq \mathbb{R}^q$ be as above. A function $f: I^q \to \mathbb{R}$ is *computable* if there is a computable sequence of rational polynomials $\{p_m(x)\}$ which converges effectively to f in uniform norm: this means there is a recursive function $e: \mathbb{N} \to \mathbb{N}$ such that for all $x \in I^q$ and all N:

$$m \geqslant e(N) \quad \text{implies} \quad |f(x) - p_m(x)| \leqslant 2^{-N}.$$

The equivalence of Definitions A and B will be proved in Sections 5 and 7.

We must now extend these definitions to sequences of functions $\{f_n(x)\}$ and also to unbounded domains. These extensions follow very closely the pattern laid out in passing from Section 1 (computable real numbers) to Section 2 (computable sequences of real numbers). This extension process is routine: whenever a new parameter enters into a definition, the dependence on this parameter must be recursive. Now for the details.

We begin with the case of a sequence of functions, still restricted to a compact domain. The corresponding extensions of Definitions A and B above are:

Definition A′ (Effective evaluation). Let $I^q \subseteq \mathbb{R}^q$ be a computable rectangle. A sequence of functions $f_n: I^q \to \mathbb{R}$ is *computable* (as a sequence) if:

(i′) for any computable sequence of points $x_k \in I^q$, the double sequence of reals $\{f_n(x_k)\}$ is computable;

(ii′) there exists a recursive function $d: \mathbb{N} \times \mathbb{N} \to \mathbb{N}$ such that for all $x, y \in I^q$ and all n, N:

$$|x - y| \leqslant 1/d(n, N) \quad \text{implies} \quad |f_n(x) - f_n(y)| \leqslant 2^{-N}.$$

Definition B′ (Effective Weierstrass). Let $I^q \subseteq \mathbb{R}^q$ be a above. A sequence of functions $f_n: I^q \to \mathbb{R}$ is *computable* (as a sequence) if there is a computable double sequence of rational polynomials $\{p_{nm}(x)\}$ with the following property. There is a recursive function $e: \mathbb{N} \times \mathbb{N} \to \mathbb{N}$ such that for all $x \in I^q$ and all n, N:

$$m \geqslant e(n, N) \quad \text{implies} \quad |f_n(x) - p_{nm}(x)| \leqslant 2^{-N}.$$

3. Computable Functions of One or Several Real Variables

We now pass from the compact rectangle I^q to the unbounded domain $\mathbb{R}^q$. This is done via a sequence of rectangles

$$I_M^q = \{-M \leq x_i \leq M, 1 \leq i \leq q\},$$

where $M = 0, 1, 2, \ldots$. The idea is to require uniform continuity or convergence on each rectangle I_M^q, effectively in M. Then the definitions parallel Definitions A' and B' above, except that we have one new parameter, M.

Definition A'' (Effective evaluation). A sequence of functions $f_n: \mathbb{R}^q \to \mathbb{R}$ is *computable (as a sequence)* if:

(i'') for any computable sequence of points $x_k \in \mathbb{R}^q$, the double sequence of reals $\{f_n(x_k)\}$ is computable;

(ii'') there exists a recursive function $d: \mathbb{N} \times \mathbb{N} \times \mathbb{N} \to \mathbb{N}$ such that for all M, n, N:

$$|x - y| \leq 1/d(M, n, N) \quad \text{implies} \quad |f_n(x) - f_n(y)| \leq 2^{-N} \quad \text{for all } x, y \in I_M^q,$$

where $I_M^q = \{-M \leq x_i \leq M, 1 \leq i \leq q\}$.

Definition B'' (Effective Weierstrass). A sequence of functions $f_n: \mathbb{R}^q \to \mathbb{R}$ is *computable (as a sequence)* if there is a computable triple sequence of rational polynomials $\{p_{Mnm}\}$ with the following property. There is a recursive function $e: \mathbb{N} \times \mathbb{N} \times \mathbb{N} \to \mathbb{N}$ such that for all M, n, N:

$$m \geq e(M, n, N) \quad \text{implies} \quad |f_n(x) - p_{Mnm}(x)| \leq 2^{-N} \quad \text{for all } x \in I_M^q,$$

where $I_M^q = \{-M \leq x_i \leq M, 1 \leq i \leq q\}$.

Remark. Without loss of generality, we can assume that the functions $d(\ )$ and $e(\ )$ are increasing in all variables. We shall frequently make this assumption.

Of course, Definitions A'' and B'' contain Definitions A' and B' (when we hold M constant); and these in turn contain Definitions A and B (when we hold n constant).

A complex-valued function is called *computable* if its real and imaginary parts are computable. Similarly for sequences of complex functions.

Standard functions. It is trivial to verify that most of the specific continuous functions encountered in analysis—e.g. e^x, $\sin x$, $\cos x$, $\log x$, $J_0(x)$ and $\Gamma(x)$, as well as $x \pm y$, xy and x/y—are computable over any computable rectangle on which they are continuous. Those without singularities are computable over $\mathbb{R}^q$ ($q = 1, 2, \ldots$) as well.

To consider whether functions like $1/x$ or $\log x$ are computable on the open interval $(0, \infty)$, we need a slight extension of our previous definitions. This we now give.

Computability on $(0, \infty)$. To define this, we simply mimic Definition A''. In fact, to define the notion "$\{f_n\}$ is computable on $(0, \infty)$", we make precisely three changes in Definition A″. First we set the dimension $q = 1$. Second, in condition (i″) we replace $x_k \in \mathbb{R}$ by $x_k > 0$. Third, in condition (ii″), we replace the domain $I_M = [-M, M]$ by the interval $[1/M, M]$. Otherwise, the definition reads exactly as before.

It is easy to verify that $1/x$ is computable on $(0, \infty)$—and also in a like manner on $(-\infty, 0)$. Also $\log x$ is computable on $(0, \infty)$.

A similar definition would apply to any open interval (a, b) with computable endpoints.

Definition B″ could also be extended in a similar manner.

A technical remark. In Definitions B, B′, and B″ above, we have used computable sequences of *rational* polynomials. In an analogous way we could define a computable sequence of *real* polynomials. The Definitions B, B′, B″ could have been given in terms of real polynomials. For, by the methods of Sections 1 and 2, the real coefficients of the real polynomials can be effectively approximated by rationals. Henceforth, in applying Definitions B, B′, B″, the versions based on real polynomials and rational polynomials will be used interchangeably.

Besides the definitions of Lacombe [1955a, 1955b] and Grzegorczyk [1955, 1957] cited above, we mention that Grzegorczyk in the same papers gave several equivalent definitions. Another equivalent definition has recently been suggested by Mycielski (cf. Pour-El, Richards [1983a]).

Finally, if we isolate condition (i) from Definition A (sequential computability), we obtain the Banach-Mazur definition. This is strictly broader than the definitions given above. An example of a sequentially computable continuous function which is not computable will be given in Chapter 1, Section 3.

4. Preliminary Constructs in Analysis

We turn now to the computable theory of functions of a real variable.

In this section we deal with composition, patching of functions, and extension of functions (Theorems 1–3 respectively). These are preliminary theorems which the usual practice of analysis takes for granted.

In Theorems 1, 1a, 1b we begin with a single computable function on a compact rectangle, and then gradually extend first to a noncompact domain, and then to a sequence of functions. These extensions are routine. Often in similar situations this is the case. However not always. In fact, sometimes the extensions do not hold at all. See, for example, Section 6 in this chapter and Sections 1 and 2 in Chapter 1.

In this section we give the proofs in all of their boring detail. Beginning in Section 5, we adopt a more compressed style.

Finally a technical note. Until the equivalence of Definitions A and B is established, we will use Definition A (or its extensions A′ and A″). Thus Theorems 1–5 in this and the next section are all based on Definition A.

Theorem 1 (Composition). *Let $g_1, \ldots, g_p$ be computable functions from $I^q \to \mathbb{R}$, and suppose that the range of the vector $(g_1, \ldots, g_p)$ is contained in a computable rectangle*

4. Preliminary Constructs in Analysis

I^p. Let $f: I^p \to \mathbb{R}$ be computable. Then the composition $f(g_1, \ldots, g_p)$ is a computable function from I^q into $\mathbb{R}$.

Proof. By Definition A, we must prove (i) sequential computability, and (ii) effective uniform continuity. For convenience, we write $\vec{g} = (g_1, \ldots, g_p)$.

Proof of (i). Let $\{x_n\}$ be a computable sequence of points in I^q. Since the functions g_i are sequentially computable, $\{\vec{g}(x_n)\}$ is a computable sequence of points in I^q. Then, since f is sequentially computable $\{f(\vec{g}(x_n))\}$ is a computable sequence of reals, as desired.

Proof of (ii). Since this is our first proof dealing with effective uniform continuity, we shall give it in great detail.

From the effective uniform continuity of $g_1, \ldots, g_p$ and f, we have recursive functions $d_1, \ldots, d_p$ and d^* such that, for all x, y in I^q or I^p, and for all N:

$$|x - y| \leq 1/d_i(N) \quad \text{implies} \quad |g_i(x) - g_i(y)| \leq 2^{-N},$$
$$|x - y| \leq 1/d^*(N) \quad \text{implies} \quad |f(x) - f(y)| \leq 2^{-N},$$

where $|\ |$ denotes the euclidean norm. We need to construct a recursive function d^{**} such that:

$$|x - y| \leq 1/d^{**}(N) \quad \text{implies} \quad |f(\vec{g}(x)) - f(\vec{g}(y))| \leq 2^{-N}.$$

We begin with a heuristic approach. Starting with the desired inequality

$$|f(\vec{g}(x)) - f(\vec{g}(y))| \leq 2^{-N} \tag{$*$}$$

we will work backwards to find a suitable $d^{**}(N)$. Then we will show that d^{**} is recursive, and that it does what is required.

By definition of d^*, $(*)$ will hold provided that

$$|\vec{g}(x) - \vec{g}(y)| \leq 1/d^*(N).$$

Since the vector $\vec{g}$ is p-dimensional, it suffices that for each component g_i,

$$|g_i(x) - g_i(y)| \leq (1/p)(1/d^*(N)).$$

A stronger (and therefore sufficient) condition is

$$|g_i(x) - g_i(y)| \leq 2^{-Q} \quad \text{where} \quad 2^Q \geq p \cdot d^*(N).$$

Now by definition of the d_i,

$$|x - y| \leq 1/d_i(Q) \quad \text{implies} \quad |g_i(x) - g_i(y)| \leq 2^{-Q}.$$

Comparing the last two displayed formulas, we are led to the following definition.

Define the recursive function $Q = Q(N)$ to be the least integer such that

$$2^Q \geq p \cdot d^*(N).$$

Then define the recursive function d^{**} by

$$d^{**}(N) = \max_{1 \leq i \leq p} d_i(Q(N)).$$

Now we verify that d^{**} serves as a modulus of continuity for $f(\vec{g})$. For, retracing our previous steps:

If $|x - y| \leq 1/d^{**}(N)$, then $|x - y| \leq 1/d_i(Q(N))$ for all i, whence $|g_i(x) - g_i(y)| \leq 2^{-Q(N)} \leq (1/p)(1/d^*(N))$ for all i, whence $|\vec{g}(x) - \vec{g}(y)| \leq 1/d^*(N)$, whence $|f(\vec{g}(x)) - f(\vec{g}(y))| \leq 2^{-N}$, as desired. $\square$

We turn now to the extension of Theorem 1 to unbounded domains. The proof has one amusing twist. Although the definition of computability involves sequential computability and effective uniform continuity, the key issue in the proof below turns out to be the *rate of growth* of the $g_i(x)$ as $|x| \to \infty$.

Theorem 1a (Composition on unbounded domains). *Let $g_1, \ldots, g_p$ be computable functions from $\mathbb{R}^q \to \mathbb{R}$, and let $f: \mathbb{R}^p \to \mathbb{R}$ be computable. Then $f(g_1, \ldots, g_p)$ is computable.*

Proof. The proof of (i) sequential computability is the same as in Theorem 1 above.

Before proving (ii), we need the following. Recall that I_M^q denotes the cube in $\mathbb{R}^q$ given by $|x_i| \leq M$, $1 \leq i \leq q$.

Lemma (Rate of growth). *Let $g: \mathbb{R}^q \to \mathbb{R}$ be a computable function. Then there exists a recursive function $a: \mathbb{N} \to \mathbb{N}$ such that for all natural numbers M:*

$$x \in I_M^q \quad \text{implies} \quad |g(x)| \leq a(M).$$

Proof of lemma. We use the fact that g is effectively uniformly continuous on each I_M^q, in a manner which varies effectively in M. More precisely, there is a recursive function $d: \mathbb{N} \times \mathbb{N} \to \mathbb{N}$ such that

$$|x - y| \leq 1/d(M, N) \quad \text{implies} \quad |g(x) - g(y)| \leq 2^{-N} \qquad \text{for all } x, y \in I_M^q.$$

In particular, setting $N = 0$

$$|x - y| \leq 1/d(M, 0) \quad \text{implies} \quad |g(x) - g(y)| \leq 1.$$

Now any point $x \in I_M^q$ can be connected to 0 by a straight line of length $\leq qM$. Suppose we break this line into $q \cdot M \cdot d(M, 0)$ equal segments, of length $\leq 1/d(M, 0)$. Then on each segment, $|g(u) - g(v)| \leq 1$. Since there are $q \cdot M \cdot d(M, 0)$ segments

4. Preliminary Constructs in Analysis

between the points x and 0, we have

$$|g(x) - g(0)| \leq q \cdot M \cdot d(M, 0) \quad \text{for all } x \in I_M^q.$$

Let C be an integer with $|g(0)| \leq C$. Define the recursive function a by

$$a(M) = q \cdot M \cdot d(M, 0) + C.$$

Then $|g(x)| \leq a(M)$ for all $x \in I_M^q$, as desired. □

Proof of (ii). Here the moduli of uniform continuity depend on the domain I_M^q, this dependence being effectively given by the recursive functions $d_i(M, N)$ and $d^*(M, N)$ (where $d_1, \ldots, d_p$ and d^* correspond to $g_1, \ldots, g_p$ and f as in Theorem 1 above). Without loss of generality, we can assume that $d_1, \ldots, d_p$ and d^* are increasing functions. The point of the Lemma is that the function $\vec{g} = (g_1, \ldots, g_p)$ maps I_M^q into a cube $I_{a(M)}^p$ whose size is a recursive function of M. From here on, the proof is almost identical with that of Theorem 1.

Let $a_i(M)$ be the function corresponding to g_i via the Lemma, and let $a(M) = \max\{a_i(M): 1 \leq i \leq p\}$.

Let $Q = Q(M, N)$ be the least integer such that

$$2^Q \geq p \cdot d^*(a(M), N).$$

Define d^{**} by

$$d^{**}(M, N) = \max_{1 \leq i \leq p} d_i(M, Q(M, N)).$$

Clearly d^{**} is recursive. The fact that, for $x, y \in I_M^q$, $|x - y| \leq 1/d^{**}(M, N)$ implies $|f(\vec{g}(x)) - f(\vec{g}(y))| \leq 2^{-N}$, is proved exactly as in Theorem 1 above. □

Now we move on to sequences of functions. We continue to use the vector notation $\vec{g} = (g_1, \ldots, g_p)$ for mappings into $\mathbb{R}^p$.

Theorem 1b. *Let $\vec{g}_n: \mathbb{R}^q \to \mathbb{R}^p$ be a computable sequence of vector-valued functions, and let $f_m: \mathbb{R}^p \to \mathbb{R}$ be a computable sequence of real functions. Then the compositions $f_m(\vec{g}_n)$ form a computable double sequence of real-valued functions.*

Proof. The construction is identical to that in Theorem 1a, except that, where appropriate, we insert the parameters m and n. Thus the functions $d_1, \ldots, d_p$ for $\vec{g}_n$ become $d_i(M, n, N)$. The function d^* for f_m becomes $d^*(M, m, N)$. The function a in the Lemma becomes $a(M, n)$. Again we can assume that these functions are increasing. Finally the d^{**} for $f_m(\vec{g}_n)$ becomes

$$d^{**}(M, m, n, N) = \max_{1 \leq i \leq p} d_i(M, n, Q(M, m, n, N)),$$

where Q is the least integer such that

$$2^Q \geq p \cdot d^*(a(M, n), m, N). \qquad \square$$

In the preceding proofs, effective uniform continuity played the key role. In the next result, we will find that sequential computability is the key issue. This result also suggests one reason why we require our intervals to have computable endpoints.

Theorem 2 (Patching theorem). *Let $[a, b]$ and $[b, c]$ be intervals with computable endpoints, and let f and g be computable functions defined on $[a, b]$ and $[b, c]$ respectively, with $f(b) = g(b)$. Then the common extension of f and g is a computable function on $[a, c]$.*

Proof. Write the common extension as $f \cup g$. Since f and g are effectively uniformly continuous, clearly $f \cup g$ is also effectively uniformly continuous. The sequential computability of $f \cup g$ is a little harder to prove.

Take any computable sequence $x_n \in [a, c]$; we need to show that the sequence $(f \cup g)(x_n)$ is computable. The difficulty is that there is no effective method for deciding in general whether $x_n < b$, $x_n = b$, or $x_n > b$. Thus we have no effective method for deciding which of the functions, f or g, should be applied to x_n.

To remedy this, we shall construct a computable double sequence $\{x_{nN}\}$ as follows. Since $\{x_n\}$ is computable, there is a computable double sequence of rationals r_{nN} such that $|x_n - r_{nN}| \leq 2^{-N}$ for all n, N. Since b is computable, there is a computable sequence of rationals s_N such that $|b - s_N| \leq 2^{-N}$ for all N.

Now computations with rational numbers can be performed *exactly*, so the following is an effective procedure. Define:

$$x_{nN} = \begin{cases} x_n & |r_{nN} - s_N| > 3 \cdot 2^{-N}, \\ x_n - 6 \cdot 2^{-N} & \text{otherwise.} \end{cases}$$

[By deleting finitely many N, we can assume that $12 \cdot 2^{-N} < (b - a)$, so that $x_{nN} \in [a, c]$ in all cases.]

Now $x_{nN} \neq b$ for all n, N, and the above inequalities furnish an effective method for deciding whether $x_{nN} < b$ or $x_{nN} > b$. Namely: $x_{nN} > b$ if $r_{nN} - s_N > 3 \cdot 2^{-N}$, and $x_{nN} < b$ otherwise.

Hence the double sequence $(f \cup g)(x_{nN})$ is computable, since we have a method which is effective in n and N for deciding whether to apply f to x_{nN} (if $x_{nN} < b$), or to apply g to x_{nN} (if $x_{nN} > b$).

Now $|x_n - x_{nN}| \leq 6 \cdot 2^{-N}$, so $x_{nN} \to x_n$ as $N \to \infty$, effectively in N and n. Since $f \cup g$ is effectively uniformly continuous, $(f \cup g)(x_{nN}) \to (f \cup g)(x_n)$ as $N \to \infty$, effectively in N and n.

We apply Proposition 1 of Section 2:

Since $(f \cup g)(x_{nN})$ is computable and converges to $(f \cup g)(x_n)$ as $N \to \infty$, effectively in all variables, $(f \cup g)(x_n)$ is computable, as desired. $\square$

5. Basic Constructs of Analysis

The patching theorem has an obvious extension to q dimensions.

Frequently in analytic arguments, one needs to extend the domain of definition of a function from a rectangle I^q to a larger rectangle I_M^q. This is done, so to speak, to give us "room around the edges". The following theorem justifies this procedure.

Theorem 3 (Expansion theorem). *Let $I^q = \{a_i \leqslant x_i \leqslant b_i, 1 \leqslant i \leqslant q\}$ be a computable rectangle in $\mathbb{R}^q$, let M be an integer, and let $I_M^q = \{-M \leqslant x_i \leqslant M, \text{ all } i\}$. Suppose that the rectangle I_M^q contains I^q in its interior. Then any computable function f on I^q can be extended to a computable function on I_M^q.*

Proof. We prove this as a corollary of Theorems 1 and 2. For $1 \leqslant i \leqslant q$, let

$$y_i(x_i) = \begin{cases} a_i & \text{for } -M \leqslant x_i \leqslant a_i, \\ x_i & \text{for } a_i \leqslant x_i \leqslant b_i, \\ b_i & \text{for } b_i \leqslant x_i \leqslant M. \end{cases}$$

Then by the patching theorem, Theorem 2, the functions y_i are computable. Furthermore, the vector $\vec{y} = (y_1, \ldots, y_q)$ maps I_M^q into the smaller rectangle I^q.

Finally, by the composition theorem, Theorem 1, the function

$$f(y_1(x_1), \ldots, y_q(x_q))$$

is computable on I_M^q. This gives the desired extension of f. □

5. Basic Constructs of Analysis

This section contains three main topics. The first is closure under effective uniform convergence (Theorem 4). The second is the equivalence of Definitions A and B (Theorem 6). The third is integration.

A preliminary result about integration (Theorem 5) is given because it is needed for the proof of Theorem 6. Once we have Theorem 6, a variety of more difficult integration theorems can be proved (Corollaries 6a, b, c). A thoroughgoing treatment of integration has been included because several types of integrals—integrals depending on a parameter, line integrals, surface integrals, etc.—occur routinely in analysis and its applications to physics. For example, in this book we use Kirchhoff's solution formula for the wave equation, which depends on integration over a sphere (cf. Chapter 3, Section 5).

[Differentiation is treated in Chapter 1; cf. the remark at the end of this section.]

We recall that, until we have the equivalence of Definitions A and B, our working definition of "computable function of a real variable" is Definition A.

We turn now to closure under effective uniform convergence. First we must define what it means for a sequence of functions to be effectively uniformly convergent. Although the definition of this term is implicit in Sections 2 and 3, we set it down:

Definition. Let $\{f_{nk}\}$ and $\{f_n\}$ be respectively a double sequence and a sequence of functions from I^q into $\mathbb{R}$. We say that $f_{nk} \to f_n$ as $k \to \infty$, *uniformly in x, effectively in k and n*, if there is a recursive function $e\colon \mathbb{N} \times \mathbb{N} \to \mathbb{N}$ such that for all n and N:

$$k \geqslant e(n, N) \quad \text{implies} \quad |f_{nk}(x) - f_n(x)| \leqslant 2^{-N} \quad \text{for all } x \in I^q.$$

Theorem 4 (Closure under effective uniform convergence). *Let $f_{nk}\colon I^q \to \mathbb{R}$ be a computable double sequence of functions such that $f_{nk} \to f_n$ as $k \to \infty$, uniformly in x, effectively in k and n. Then $\{f_n\}$ is a computable sequence of functions.*

Proof. For simplicity, we suppress the index n, and consider a sequence of functions f_k which is effectively uniformly convergent to a limit f. The extension to $f_{nk} \to f_n$ is left to the reader. We begin by proving (ii) that f is effectively uniformly continuous.
By hypothesis, there is a recursive function $e\colon \mathbb{N} \to \mathbb{N}$ such that

$$k \geqslant e(N) \quad \text{imples} \quad |f_k(x) - f(x)| \leqslant 2^{-N} \quad \text{for all } x \in I^q.$$

Since the sequence $\{f_k\}$ is uniformly continuous, effectively in k, there is a recursive function $d\colon \mathbb{N} \times \mathbb{N} \to \mathbb{N}$ such that for all $x, y \in I^q$ and all k and N:

$$|x - y| \leqslant 1/d(k, N) \quad \text{implies} \quad |f_k(x) - f_k(y)| \leqslant 2^{-N}.$$

Replacing N by $N + 2$ throughout, the error 2^{-N} becomes $2^{-N-2} = (1/4) \cdot 2^{-N} < (1/3) \cdot 2^{-N}$. Then to bound $|f(x) - f(y)|$, we compare $f(x)$ to $f_k(x)$ to $f_k(y)$ to $f(y)$ in the standard way. Thus we define the recursive function d^* by:

$$d^*(N) = d(e(N + 2), N + 2).$$

We now show that d^* serves as an effective modulus of continuity for f. Take any $x, y \in I^q$ with $|x - y| \leqslant 1/d^*(N)$. Set $k = e(N + 2)$. Then $|f(x) - f_k(x)| < (1/3) \cdot 2^{-N}$ (by definition of e); $|f_k(x) - f_k(y)| < (1/3) \cdot 2^{-N}$ (by definition of d); and $|f_k(y) - f(y)| < (1/3) \cdot 2^{-N}$ (by definition of e). Hence $|f(x) - f(y)| \leqslant 2^{-N}$, as desired.

Proof of (i) (sequential computability). There is one amusing point in this otherwise routine proof. Take any computable sequence $\{x_m\}$ in I^q; we need to show that $\{f(x_m)\}$ is computable. Since $f_k(x_m) \to f(x_m)$ as $k \to \infty$, it would suffice, in view of Proposition 1 in Section 2, if the convergence were effective in k and m. But in fact we have more: since $f_k(x) \to f(x)$ as $k \to \infty$, effectively in k and uniformly in x, the convergence of $f_k(x_m)$ to $f(x_m)$ is actually *effective in k and uniform in m*.
(Thus the weaker condition of effectiveness = "logical uniformity" is here deduced from uniformity in the analytic sense—a circumstance which is rather rare.) □

5. Basic Constructs of Analysis

We turn now to a preliminary result on the computability of integrals. This is needed to prove the equivalence of Definitions A and B (Theorem 6 below). Once we have Theorem 6, we shall find that the result on integration can easily be extended to a much more general setting.

Theorem 5 (Definite Integrals). *Let I^q be a computable rectangle in $\mathbb{R}^q$, and let $f_n: I^q \to \mathbb{R}$ be a computable sequence of functions. Then the definite integrals*

$$v_n = \int \cdots \int_{I^q} f_n(x_1, \ldots, x_q)\, dx_1 \ldots dx_q$$

form a computable sequence of real numbers.

Proof. We effectivize the Riemann sum definition of the integral. For simplicity, we take $q = 1$, leaving the general case to the reader. Thus the integration is over a 1-dimensional interval $[a, b]$, and

$$v_n = \int_a^b f_n(x)\, dx.$$

For any $k \geq 1$, let v_{nk} be the k-th Riemann sum approximation

$$v_{nk} = \frac{b-a}{k} \cdot \sum_{j=1}^{k} f_n\left(a + \frac{j}{k}(b-a)\right).$$

Since a, b are computable reals, and since $\{f_n\}$ is sequentially computable, the double sequence $\{v_{nk}\}$ is computable.

To show that v_{nk} converges to v_n, effectively in k and n, we use the effective uniform continuity of the f_n. Thus there is a recursive function $d(n, N)$ such that

$$|x - y| \leq 1/d(n, N) \quad \text{implies} \quad |f_n(x) - f_n(y)| \leq 2^{-N}.$$

Let M be an integer $> (b - a)$, and let

$$e(n, N) = M \cdot d(n, N).$$

We will show that

$$k \geq e(n, N) \quad \text{implies} \quad |v_{nk} - v_n| \leq (b - a) \cdot 2^{-N}. \tag{$*$}$$

To prove $(*)$, let I_j be the j-th subinterval corresponding to the above Riemann sum, that is

$$I_j = \left[a + \frac{j-1}{k}(b-a),\, a + \frac{j}{k}(b-a)\right].$$

Suppose $k \geqslant e(n, N)$. Then each subinterval I_j has length equal to $(b - a)/k$ where

$$(b - a)/k < M/k \leqslant M/e(n, N) = 1/d(n, N).$$

By definition of $d(n, N)$, this means that the function f_n varies by $\leqslant 2^{-N}$ over each interval I_j. Hence the difference $(v_{nk} - v_n)$ between the Riemann sum and the integral satisfies:

$$|v_{nk} - v_n| \leqslant \sum_{j=1}^{k} (\text{maximum variation of } f_n \text{ over } I_j) \cdot (\text{length of } I_j)$$

$$\leqslant k \cdot 2^{-N} \cdot \frac{(b-a)}{k}$$

$$\leqslant (b - a) \cdot 2^{-N},$$

proving (∗).

Since the function $e(n, N)$ is recursive, (∗) implies that $v_{nk} \to v_n$ as $k \to \infty$, effectively in k and n. Since $\{v_{nk}\}$ is computable and approaches v_n as $k \to \infty$, effectively in both variables, $\{v_n\}$ is computable (cf. Proposition 1 in Section 2). □

Equivalence of Definitions A and B

Up to now, all of our work has been based on Definition A and its extensions A' and A''. We now consider the equivalence of Definitions A and B. This is the content of Theorem 6 below. Since Definition B is based on polynomial approximation, this amounts to giving an effective treatment of the Weierstrass Approximation Theorem.

Actually, the proof of the Effective Weierstrass Theorem is quite complicated. We postpone it until Section 7. The reason for stating Theorem 6 here is that its corollaries (Corollaries 6a, b, c) properly belong in this section. On the other hand, the proof in Section 7 uses only Theorems 1–5.

We remark, however, that part of the proof of Theorem 6 below—the easy part—is given right here in this section. This is done so as not to clutter up the Effective Weierstrass proof in Section 7.

We recall that Definitions A and B apply to a single real-valued function f defined on a computable rectangle I^q. Definitions A' and B' apply to a sequence of functions f_n on I^q, and A''/B'' apply to a sequence of functions on $\mathbb{R}^q$.

Theorem 6 (Equivalence of Definitions A and B). *Definitions A and B are equivalent. Similarly for Definitions A' and B', and for Definitions A'' and B''.*

Proof. We shall give the proof for Definitions A and B. The extension to A'/B' and A''/B'' is routine, involving only a proliferation of indices.

Furthermore, we consider only the case of dimension $q = 1$. The extension to q dimensions is also routine.

5. Basic Constructs of Analysis

Definition B implies Definition A. Let f satisfy Definition B. Then there exists a computable sequence of polynomials $\{p_m\}$ which converges effectively and uniformly to f. By Theorem 4 above (closure under effective uniform convergence), f is computable in the sense of Definition A.

Definition A implies Definition B. The proof will be given in Section 7. □

Integration

Here the preliminary Theorem 5 above is extended to cover a variety of integration processes. These extensions depend on Theorem 6, which allows us to use Definition B in place of Definition A. In fact, the powerful and general Corollary 6c could hardly be proved in any other way.

Corollary 6c combines in one statement: (1) integrals depending on a parameter t; (2) integration over a general class of regions in $\mathbb{R}^3$ (i.e. regions—like the interior of a sphere—which are not rectangles); and (3) line integrals and surface integrals.

[Of course, instead of $\mathbb{R}^3$ we could have stated Corollary 6c for $\mathbb{R}^n$; we used $\mathbb{R}^3$ for convenience.]

In order to achieve this generality, Corollary 6c requires a bit of preface. The conditions involve compact regions K in $\mathbb{R}^3$ and measures μ on K. At first glance, these considerations may appear a trifle ponderous. In fact, however, they ae entirely natural. Consider, for example, a compact surface K in $\mathbb{R}^3$. Then K has Lebesgue measure zero. But the very notion of surface integration implies that there is some measure μ (e.g. the area measure) with respect to which we are integrating. Hence we have the pair $\langle K, \mu \rangle = \langle \text{compact set, measure} \rangle$. This is absolutely standard, and we have merely put the situation into a general format in order to cover a wide variety of applications.

The main hypothesis in Corollary 6c is that the monomials $x^a y^b z^c$ have computable integrals with respect to K and μ, effectively in a, b, and c. This is easily verified in most applications. Then the Effective Weierstrass condition of Definition B permits an immediate extension from the monomials to arbitrary computable functions.

[By contrast, a proof based on Definition A would be pretty ugly, even for such a simple domain as the surface of the unit sphere in $\mathbb{R}^3$.]

Corollaries 6a and 6b involve elementary results which seem important enough to be displayed separately.

Corollary 6a (Indefinite integrals). *Let f be a computable function on a computable interval $[a, b]$. Then the indefinite integral*

$$\int_a^x f(u)\, du$$

is computable on $[a, b]$.

Corollary 6b (Definite integrals depending on a parameter). *Let $f(x, t)$ be a computable function on the rectangle $[a, b] \times [c, d]$. Then the function*

$$F(t) = \int_a^b f(x, t)\, dx$$

is computable on $[c, d]$.

Proofs. See the proof of Corollary 6c. □

As a preparation for Corollary 6c we need:

Definition. Let K be a compact set in R^q. We say that a real-valued function f on K is *computable* if f has an extension to a computable function on a computable rectangle $I^q \supseteq K$.

For notational convenience, we give the next definition for 3-dimensional space, and write a typical point in $\mathbb{R}^3$ as (x, y, z).

Definition. Let K be a compact set in $\mathbb{R}^3$, and let μ be a finite measure on K. We say that the pair $\langle K, \mu \rangle$ is *computably integrable* if for $a, b, c \in \mathbb{N}$,

$$v_{abc} = \int_K x^a y^b z^c \, d\mu$$

is a computable triple sequence of real numbers.

Notes. We have not defined either a "computable compact set" or a "computable measure". The above definition expresses a property of the pair $\langle K, \mu \rangle$. Moreover, this definition does not require that the integrals be computed by some "constructive method"; merely that the *results* of the integration yield a computable sequence.

Examples. First, let μ be Lebesgue measure. Practically every specific region K encountered in elementary analysis is *computably integrable*. For all that is required is that the integrals of $x^a y^b z^c$ over K yield a computable sequence of values. In particular, the following are computably integrable with respect to Lebesgue measure:

A disk with computable center and radius,
An ellipsoid with computable center and axes,
A computable rectangle,
A polyhedron given by computable parameters,
A cylinder or cone given by computable parameters.

A second class of example involves line or surface integrals. One instance which is important in physical applications is:

5. Basic Constructs of Analysis

The unit sphere $K = \{x^2 + y^2 + z^2 = 1\}$, with the area measure $\mu = d\sigma$ normalized so that the total area equals 1.

The next result covers all of these cases, and also allows integrals depending on a parameter.

Corollary 6c (Integration over regions). *Let K be a compact set in $\mathbb{R}^3$, and let μ be a finite measure on K. Suppose that the pair $\langle K, \mu \rangle$ is computably integrable. Let I^q be a computable rectangle in $\mathbb{R}^q$. Let $f(x, y, z, t)$, $t \in I^q$, be computable on $K \times I^q$. Then*

$$F(t) = \iiint_K f(x, y, z, t) \, d\mu(x, y, z)$$

is computable on I^q.

Proof. We use Definition B. Since $f(x, y, z, t)$ is computable, there is a computable sequence of polynomials $\{p_m(x, y, z, t)\}$ which converges to $f(x, y, z, t)$ as $m \to \infty$, uniformly in (x, y, z, t) and effectively in m.

These polynomials are computable finite linear combinations of the monomials

$$x^a y^b z^c t^d,$$

where

$$t^d = t_1^{d_1} t_2^{d_2} \ldots t_q^{d_q}.$$

[Actually, since the integration involves only x, y, z, the t^d terms behave like constants and can be taken outside of the integral.]

Now since the pair $\langle K, \mu \rangle$ is computably integrable,

$$\iiint_K x^a y^b z^c \, d\mu(x, y, z)$$

is a computable triple sequence of real numbers.

Hence the sequence of polynomials

$$P_m(t) = \iiint p_m(x, y, z, t) \, d\mu(x, y, z)$$

is computable.

Finally, since $p_m \to f$, uniformly in (x, y, z, t) and effectively in m, and since the set K has finite μ-measure,

$$P_m(t) \to F(t) \quad \text{as } m \to \infty,$$

uniformly in t and effectively in m. Hence by Theorem 4, $F(t)$ is computable as desired. □

After such an extended study of integration, the reader may wonder why we do not consider differentiation. The reason is that the computable theory of differentiation is more difficult. It is dealt with in Chapter 1.

6. The Max-Min Theorem and the Intermediate Value Theorem

This is the first section in which the distinction between computable elements and computable sequences finds illustration in a natural setting.

The maximum value of a computable function is computable (Theorem 7). This result extends to computable sequences of functions. By contrast, the Effective Intermediate Value Theorem (Theorem 8) holds for individual computable functions— but it does not hold for computable sequences of functions (Example 8a).

In addition, this section includes the following. Theorem 9 asserts that the computable reals form a real closed field. The section closes with a few remarks about the Mean Value Theorem.

Theorem 7 (Maximum Values). *Let I^q be a computable rectangle in $\mathbb{R}^q$, and let $f_n\colon I^q \to \mathbb{R}$ be a computable sequence of functions. Then the maximum values*

$$s_n = \max\{f_n(x)\colon x \in I^q\}$$

form a computable sequence of real numbers.

Proof. As previously, we shall treat the 1-dimensional case, where the f_n are defined on a computable interval $[a, b]$, and leave the q-dimensional case to the reader.

We use Definition A. The proof, like that of Theorem 5 above (definite integrals), is based on a partitioning of the interval $[a, b]$. For any $k \geqslant 1$, let s_{nk} be the "partial maximum"

$$s_{nk} = \max\left\{f_n\left[a + \frac{j}{k}(b-a)\right]\colon 1 \leqslant j \leqslant k\right\}.$$

Since a, b are computable reals, and since $\{f_n\}$ is sequentially computable, the double sequence $\{s_{nk}\}$ is computable.

By the effective uniform continuity of $\{f_n\}$, there is a recursive function $d(n, N)$ such that

$$|x - y| \leqslant 1/d(n, N) \quad \text{implies} \quad |f_n(x) - f_n(y)| \leqslant 2^{-N}.$$

Let M be an integer $> (b - a)$, and let

$$e(n, N) = M \cdot d(n, N).$$

6. The Max-Min Theorem and the Intermediate Value Theorem

Then, as in Theorem 5 above, it follows that

$$k \geqslant e(n, N) \quad \text{implies} \quad |s_{nk} - s_n| \leqslant 2^{-N}.$$

Thus $s_{sk} \to s_n$ as $k \to \infty$, effectively in k and n. Since $\{s_{nk}\}$ is computable, this implies, by Proposition 1 in Section 2, that $\{s_n\}$ is computable. □

Remark. Although the maximum value of a computable function $f(x)$ is computable, the point(s) x where this maximum occurs need not be. Specker [1959] has given an example of a computable function f on [0, 1] which does not attain its maximum at any computable point. (For alternative constructions, see Kreisel [1958] and Lacombe [1957b].) In these examples, there are infinitely many maximum points. This is inevitable, for it can be shown:

If a computable function f takes a local maximum at an isolated point x (i.e. if $f(y) < f(x)$ for all y sufficiently close to x with $y \neq x$), then x is computable.

We do not need this result and shall not prove it.

Our next result is interesting in that it holds for single functions f and does not hold for sequences of functions.

Theorem 8 (Intermediate Value Theorem). *Let $[a, b]$ be an interval with computable endpoints, and let f be a computable function on $[a, b]$ such that $f(a) < f(b)$. Let s be a computable real with $f(a) < s < f(b)$. Then there exists a computable point c in (a, b) such that $f(c) = s$.*

Proof. We can assume without loss of generality that the domain of f is [0, 1] and that $s = 0$.

Now the proof breaks into two cases:

Case 1. There is some rational number c such that $f(c) = 0$. Then c is computable, and we are finished.

Case 2. $f(c) \neq 0$ for all rational c.

Now this assumption allows us to effectivize a procedure which would otherwise not be effective. Namely:

Consider $f(1/2)$. Since $f(1/2) \neq 0$, a sufficiently good approximation to $f(1/2)$ will allow us to decide—effectively—whether $f(1/2) > 0$ or $f(1/2) < 0$ (cf. Proposition 0 in Section 1). In the former case we replace the interval [0, 1] by [0, 1/2]; in the latter case, we replace [0, 1] by [1/2, 1].

We continue in this manner. After the mth stage, we have an interval $[a_m, b_m]$ of length $1/2^m$, with rational endpoints, and with $f(a_m) < 0$ and $f(b_m) > 0$. We take the midpoint $d_m = (a_m + b_m)/2$, and compute $f(d_m)$ with sufficient accuracy to determine whether $f(d_m) > 0$ or $f(d_m) < 0$ (again using the fact that, since d_m is rational, $f(d_m) \neq 0$).

If $f(d_m) > 0$, we set $a_{m+1} = a_m$, $b_{m+1} = d_m$.

If $f(d_m) < 0$, we set $a_{m+1} = d_m$, $b_{m+1} = b_m$.

The sequences $\{a_m\}$ and $\{b_m\}$ converge from below, and from above respectively, to a point c such that $f(c) = 0$. Since $b_m - a_m = 2^{-m}$, the convergence is effective. Hence c is computable. □

Now we give a counterexample for sequences of functions.

Example 8a (Failure of the Intermediate Value Theorem for sequences). There exists a computable sequence of functions f_n on $[0, 1]$ such that $f_n(0) = -1$ for all n, $f_n(1) = 1$ for all n, but there is no computable sequence of points c_n in $[0, 1]$ with $f_n(c_n) = 0$ for all n.

Proof. We use a recursively inseparable pair of sets A and B, i.e. disjoint subsets A and B of $\mathbb{N}$ which are recursively enumerable, but such that there is no recursive set C with $A \subseteq C$ and $B \cap C = \emptyset$.

Let $a: \mathbb{N} \to \mathbb{N}$ and $b: \mathbb{N} \to \mathbb{N}$ be recursive functions giving one-to-one listings of the sets A and B respectively.

[A rough summary of the following construction is given in the Notes at the end. We put these comments at the end since, without the details, they might be found more vague than helpful.]

Each function $f_n(x)$ will be piecewise linear and continuous, with the breaks in its derivative occurring at the points $1/3$ and $2/3$. Thus each f_n is determined by the four values $f_n(0), f_n(1/3), f_n(2/3), f_n(1)$. We set

$$f_n(0) = -1 \quad \text{for all } n$$
$$f_n(1) = 1 \quad \text{for all } n.$$

The interesting points are $1/3$ and $2/3$. Hence we set:

$$f_n(1/3) = \begin{cases} -1/2^m & \text{if } n \in A, n = a(m), \\ 0 & \text{if } n \notin A. \end{cases}$$

$$f_n(2/3) = \begin{cases} 1/2^m & \text{if } n \in B, n = b(m), \\ 0 & \text{if } n \notin B. \end{cases}$$

It is by no means clear from this description that $\{f_n\}$ is computable—for we have no effective test to determine whether $n \in A$, $n \in B$, or $n \in \mathbb{N} - (A \cup B)$. However, $\{f_n\}$ actually is computable as we now show.

We define a double sequence $\{f_{nk}\}$ of piecewise linear functions by:

$$f_{nk}(0) = -1, \quad f_{nk}(1) = 1 \quad \text{for all } n, k.$$

$$f_{nk}(1/3) = \begin{cases} -1/2^m & \text{if } n = a(m) \text{ for some } m \leq k, \\ 0 & \text{otherwise.} \end{cases}$$

$$f_{nk}(2/3) = \begin{cases} 1/2^m & \text{if } n = b(m) \text{ for some } m \leq k, \\ 0 & \text{otherwise.} \end{cases}$$

6. The Max-Min Theorem and the Intermediate Value Theorem

Now the double sequence $\{f_{nk}\}$ is computable. For, given any k, we have only to test the $2(k+1)$ values $a(0), \ldots, a(k)$ and $b(0), \ldots, b(k)$ in order to determine $f_{nk}(1/3)$ and $f_{nk}(2/3)$.

We shall show that $|f_{nk}(x) - f_n(x)| \leqslant 2^{-k}$ for all n, k, and x. To see this, we first verify that

$$|f_{nk}(1/3) - f_n(1/3)| \leqslant 2^{-k}$$
$$|f_{nk}(2/3) - f_n(2/3)| \leqslant 2^{-k}.$$

Namely, consider the point $1/3$. There are three cases:

1. $n \notin A$. Then $f_{nk}(1/3) = f_n(1/3)$ for all k.
2. $n \in A$, and $n = a(m)$ for some $m \leqslant k$. Then $f_{nk}(1/3) = f_n(1/3)$ for this k.
3. $n \in A$, and $n = a(m)$ with $m > k$. Then $|f_{nk}(1/3) - f_n(1/3)| = |0 - (-2^{-m})| = 2^{-m} < 2^{-k}$.

The point $2/3$ is handled similarly.

Finally, since all of the functions f_{nk} and f_n are piecewise linear, and determined by their values at $x = 1/3$ and $x = 2/3$, the inequalities which we have established for $x = 1/3$ or $2/3$ extend to all x.

Since $\{f_{nk}(x)\}$ is computable and converges to $\{f_n(x)\}$ as $k \to \infty$, uniformly in x and effectively in k and n, $\{f_n\}$ is computable (Theorem 4).

We still have to verify that there is no computable sequence $\{c_n\}$ with $f_n(c_n) = 0$ for all n.

Suppose otherwise. Since $\{c_n\}$ is computable, there is a computable sequence of rationals $\{r_n\}$ with

$$|r_n - c_n| \leqslant 1/12 \qquad \text{for all } n.$$

Since exact comparisons are effective for rational numbers, we can define a recursive set C by:

$$n \in C \qquad \text{if and only if } r_n \geqslant 1/2.$$

Now if $n \in A$, the only zero of $f_n(x)$ occurs at $x = 2/3$ (since $f_n(1/3)$ was depressed to a value slightly below zero). Similarly, if $n \in B$, the only zero of $f_n(x)$ occurs at $x = 1/3$. Since the distance from $1/3$ (or $2/3$) to $1/2$ is $1/6$, and since $1/6 > 1/12$, we have:

$$r_n > 1/2 \qquad \text{if } n \in A,$$
$$r_n < 1/2 \qquad \text{if } n \in B,$$

whence

$$A \subseteq C,$$
$$B \cap C = \emptyset.$$

Hence the recursive set C separates A and B, a contradiction. □

Notes. Of course, in this construction, the interesting action takes place on the interval [1/3, 2/3]. For $n \notin A \cup B$, the function $f_n(x)$ is identically zero on this subinterval. But if $n \in A, n = a(m)$, we depress the value $f_n(1/3)$ slightly—the amount of decrease, 2^{-m}, becoming less and less the longer we have to wait for m to occur.

Similarly if $n \in B, n = b(m)$, we increase $f_n(2/3)$ by 2^{-m}.

Since $2^{-m} \to 0$ effectively as $m \to \infty$, we can approximate $f_n(x)$, to any desired degree of precision, by checking only finitely many values of m. That is why the sequence $\{f_n(x)\}$ is computable.

On the other hand, by this "see-saw" construction, we can send the zeros of $f_n(x)$ shooting off in either direction—to 2/3 or 1/3—by an arbitrarily small perturbation in $f_n(x)$.

These ideas, in various guises, form the basis for many counterexamples.

One final note. In the proof of Theorem 8 above, there was a case analysis which began—Case 1: if c is rational, then we are done. Any single rational number is ipso-facto computable. But a sequence of rationals need not be.

Theorem 9 (Real closed field). *The computable reals form a real closed field. That is, if a polynomial has computable real coefficients, then all of its real roots are computable.*

Proof. For simple roots, this is an immediate consequence of the Intermediate Value Theorem, Theorem 8 above. For multiple roots we reason as follows. Let $p(x)$ be a polynomial with computable real coefficients, and let c be a root of order n of $p(x)$. Then the $(n - 1)$st derivative $p^{(n-1)}(x)$ has computable real coefficients, and c is a simple root of $p^{(n-1)}(x)$. Hence the previous argument applies. □

Remark (The Mean Value Theorem). In connection with the topics of this section, the reader may wonder why we do not deal with the Mean Value Theorem. The facts are these. The Mean Value Theorem effectivizes for an individual computable function with a continuous derivative (computable or not). But the theorem does not effectivize for computable sequences of functions, even if the sequence of derivatives is computable.

We have omitted this theorem for two reasons. First, we have no need for it. Second, it is hard to see how an effective version of the Mean Value Theorem could be useful. Consider the situation in classical (noncomputable) analysis. When one uses the Mean Value Theorem—$f(b) - f(a) = (b - a)f'(\xi)$ for some ξ—it is only the existence of ξ which is relevant, and not its actual value. In fact, in almost all applications, the Mean Value Theorem is used to establish inequalities involving $f(b) - f(a)$, and the location of the point ξ is immaterial.

7. Proof of the Effective Weierstrass Theorem

With this result, we prove the difficult half of Theorem 6 (equivalence of Definitions A and B). Namely, we proved in Section 5 that, if a function f satisfies the conditions of Definition B (Effective Weierstrass), then it satisfies Definition A (Effective evalua-

7. Proof of the Effective Weierstrass Theorem

tion). This proof was quite easy. However, the converse, which involves an effectivization of the classical Weierstrass Approximation Theorem, is considerably more complicated. We now prove that converse.

Here is where we get our hands dirty. This is the only effectivization of a major classical proof given in this book. It seems worthwhile to do such a thing once, if only to show what such "effectivization proofs" look like. As stated earlier, we prefer to develop general theorems (which are usually theorems with no classical analog), and deduce results like the Weierstrass theorem as corollaries.

A much simpler proof of the Effective Weierstrass Theorem is given in Chapter 2, Section 5. That proof is based on the "axioms for computability on a Banach space" developed there. Naturally one must ask—what elementary results are needed to validate the axioms—in order to make sure that no circularity has crept in. As we shall see in Chaper 2, the verification of the axioms requires Theorems 1, 4, and 7 of this chapter—none of which depend on the Effective Weierstrass Theorem.

For convenience we restate the theorem. This theorem has no number, since it is really a part of Theorem 6.

Theorem (Effective Weierstrass Theorem). *Let $[a, b]$ be an interval with computable end points, and let f be a function on $[a, b]$ which is computable in the sense of Definition A. Then there exists a computable sequence of polynomials $\{p_m\}$ which converges effectively and uniformly to f on $[a, b]$—i.e. f is computable in the sense of Definition B.*

Technical preliminaries for the effective Weierstrass proof. We select an integer M such that:

$$[a, b] \subseteq [-M/4, M/4].$$

Then we use the Expansion Theorem (Theorem 3) to extend f from $[a, b]$ to $[-M, M]$. We define a polynomial pulse function $P_m(x)$ on $[-M, M]$ by:

$$P_m(x) = \left[1 - \left(\frac{x}{M}\right)^2\right]^m.$$

We must investigate the behavior of this pulse as $m \to \infty$.

Fix a small interval $[-1/d, 1/d]$ about 0. Here d is a positive integer with $1/d < M/4$. Let J be the complementary domain

$$J = [-M, M] - [-1/d, 1/d].$$

It seems apparent that, for large m, most of the mass of the pulse $P_m(x)$ should be concentrated on $[-1/d, 1/d]$. Equivalently, only a negligible portion of the mass should lie on the complementary domain J. To make this effective, we need some explicit inequalities.

For the domain J we have:

$$|x| \geq 1/d \quad \text{implies} \quad P_m(x) \leq [1 - (1/dM)^2]^m.$$

Now consider the *smaller* interval $[-1/2d, 1/2d]$. On this interval:

$$|x| \leq 1/2d \quad \text{implies} \quad P_m(x) \geq [1 - (1/2dM)^2]^m.$$

Thus we are led to consider the ratio:

$$[1 - (1/2dM)^2]^m / [1 - (1/dM)^2]^m.$$

It is obvious that, since d and M are fixed, this ratio grows without bound as $m \to \infty$. Specifically:

$$\left[\frac{1 - (1/2dM)^2}{1 - (1/dM)^2} \right]^m \geq \frac{3m}{4d^2 M^2}. \tag{$*$}$$

For completeness we give the elementary proof. The left side of $(*)$ is just $[(1 - u)/(1 - 4u)]^m$, where $u = (1/2dM)^2$.

By induction on k, $(1 + u)^k \geq 1 + ku$ and $(1 - u)^k \geq 1 - ku$ for $k = 0, 1, 2, \ldots$. We simply apply these facts several times:

Since $(1 - u)^4 \geq 1 - 4u$, we have $(1 - u)/(1 - 4u) \geq 1/(1 - u)^3$. Also $1/(1 - u)^3 \geq (1 + u)^3$ (since $(1 - u)(1 + u) \leq 1$), and $(1 + u)^3 \geq 1 + 3u$. Finally, $(1 + 3u)^m \geq 1 + 3mu \geq 3mu$. Thus $[(1 - u)/(1 - 4u)]^m \geq 3mu$, which is $(*)$.

The main point of this subsection is contained in the following:

Technical Lemma. *Let J denote the complement of the interval $[-1/d, 1/d]$ in $[-M, M]$. Then*

$$\frac{\int_J P_m(x)\, dx}{\int_{-M}^{M} P_m(x)\, dx} \leq \frac{8d^3 M^3}{3m}.$$

Proof. Since $P_m(x) > 0$ on $(-M, M)$, the integral of P_m over $[-M, M]$ exceeds the integral over $[-1/2d, 1/2d]$. Therefore it suffices to estimate the *larger* ratio

$$\left(\int_J P_m \right) \Big/ \left(\int_{-1/2d}^{1/2d} P_m \right).$$

Now we use $(*)$ above. Thus the ratio

$$\frac{\max \text{ of } P_m(x) \text{ on } J}{\min \text{ of } P_m(x) \text{ on } [-1/2d, 1/2d]} \leq \frac{4d^2 M^2}{3m}.$$

On the other hand, the ratio

$$\frac{\text{length of } J}{\text{length of } [-1/2d, 1/2d]} \leq 2dM.$$

Multiplying the last two displayed inequalities gives the lemma. □

7. Proof of the Effective Weierstrass Theorem

Proof of the Effective Weierstrass Theorem, completed. Let $P_m(x)$ be the polynomial pulse function as above, and define

$$p_m(x) = \frac{1}{C_m} \int_{-M/2}^{M/2} P_m(t-x) f(t) \, dt, \quad -M/4 \leq x \leq M/4,$$

where

$$C_m = \int_{-M}^{M} P_m(x) \, dx.$$

Recall that the interval $[-M/4, M/4]$ contains $[a, b]$.

We now show that $\{p_m\}$ is a computable sequence of polynomials.

Clearly the sequence $P_m(x) = [1 - (x/M)^2]^m$ is computable, and by the Binomial Theorem,

$$P_m(t-x) = \sum_{k=0}^{2m} \sum_{j=0}^{k} b_{mkj} x^j t^{k-j},$$

where the triple sequence $\{b_{mkj}\}$ is computable. Then

$$p_m(x) = \frac{1}{C_m} \sum_{k=0}^{2m} \sum_{j=0}^{k} b_{mkj} \left(\int_{-M/2}^{M/2} t^{k-j} f(t) \, dt \right) \cdot x^j$$

is a computable sequence of polynomials:

For we already have that $\{b_{mkj}\}$ is computable, and the computability of C_m and of the other definite integrals above, effectively over m, k, j, follows from Theorem 5.

Now, to complete the proof, we must show that $p_m(x) \to f(x)$, uniformly in x and effectively in m, as $m \to \infty$.

Since f is effectively uniformly continuous, there is a recursive function $d(N)$ such that

$$|x - y| \leq 1/d(N) \quad \text{implies} \quad |f(x) - f(y)| \leq \frac{2^{-N}}{3}.$$

Also, since f is continuous on $[-M, M]$, there exists an integer S with

$$\sup_{|x| \leq M} |f(x)| \leq S.$$

FIX a point $x \in [-M/4, M/4]$. We have to bound $|p_m(x) - f(x)|$ in a manner uniform in x and effective in m. Since $f(x)$ is held constant, and $C_m = \int_{-M}^{M} P_m(t) \, dt$,

$$f(x) = \frac{1}{C_m} \int_{x-M}^{x+M} P_m(t-x) f(x) \, dt.$$

On the other hand,

$$p_m(x) = \frac{1}{C_m} \int_{-M/2}^{M/2} P_m(t - x) f(t) \, dt.$$

For simplicity, we write d for $d(N)$. We break the difference $p_m(x) - f(x)$ into three parts:

$$p_m(x) - f(x) = (A) + (B) + (C),$$

where

$$(A) = \frac{1}{C_m} \left[\int_{-M/2}^{M/2} P_m(t - x) f(t) \, dt - \int_{x-1/d}^{x+1/d} P_m(t - x) f(t) \, dt \right],$$

$$(B) = \frac{1}{C_m} \left[\int_{x-1/d}^{x+1/d} P_m(t - x) [f(t) - f(x)] \, dt \right],$$

$$(C) = \frac{1}{C_m} \left[\int_{x-1/d}^{x+1/d} P_m(t - x) f(x) \, dt - \int_{x-M}^{x+M} P_m(t - x) f(x) \, dt \right].$$

Concerning the various domains of integration: Of course, $t \in [x - M, x + M]$ puts $(t - x) \in [-M, M]$, so that the pulse $P_m(t - x)$ is integrated over the correct set. Since $|x| \leq M/4$, the interval $[x - M, x + M]$ contains $[-M/2, M/2]$. Also, the small interval $[x - 1/d, x + 1/d]$ lies inside $[-M/2, M/2]$. Finally, and most importantly:

In both (A) and (C), the domain of integration satisfies $(t - x) \in J$, where $J = [-M, M] - [-1/d, 1/d]$ as in the Technical Lemma above.

We recall that $C_m = \int_{-M}^{M} P_m$, and that S dominates the sup of $|f(x)|$. Hence by the Technical Lemma:

$$|(A)| \leq S \cdot \frac{8d^3 M^3}{3m},$$

$$|(C)| \leq S \cdot \frac{8d^3 M^3}{3m}.$$

To estimate (B), we use the definition of $d = d(N)$. Since $|f(t) - f(x)| \leq 2^{-N}/3$ for $t \in [x - 1/d, x + 1/d]$, and since C_m dominates the integral of $P_m(t - x)$,

$$|(B)| \leq \tfrac{1}{3} \cdot 2^{-N}.$$

Now to effectivize the bounds on (A), (B), and (C), we define the recursive function

$$m(N) = 8SM^3 \cdot d(N)^3 \cdot 2^N,$$

7. Proof of the Effective Weierstrass Theorem

so that

$$S \cdot \frac{8M^3 \cdot d(N)^3}{3 \cdot m(N)} = \frac{1}{3} \cdot 2^{-N}.$$

Then $m \geqslant m(N)$ gives

$$|(A)| \leqslant \tfrac{1}{3} \cdot 2^{-N} \quad \text{and} \quad |(C)| \leqslant \tfrac{1}{3} \cdot 2^{-N},$$

whence

$$|p_m(x) - f(x)| \leqslant |(A) + (B) + (C)| \leqslant 2^{-N}.$$

Since $m(N)$ is a recursive function, the last inequality above shows $p_m(x) \to f(x)$ effectively in m. None of the above inequalities depend on x, so the convergence is uniform in x as well. This proves the Effective Weierstrass Theorem. □

Chapter 1
Further Topics in Computable Analysis

Introduction

This chapter starts where Chapter 0 leaves off, but involves a transition to more difficult topics. It begins with differentiation and the classes C^n, $1 \leq n \leq \infty$, and progresses through the computable theory of analytic functions. It concludes with a general theorem on "translation invariant operators" which subsumes several of our previous results. We continue to use the "Chapter 0" notion of computability for continuous functions.

Naively, one might suppose that since the derivative, the operations of complex analysis, etc. are given by formulas, they should be computable. Of course, as we know, this is not necessarily the case. In fact, it is not always easy to guess what the results should be. Consider, for example, differentiation, the topic of Section 1. If a computable function possesses a continuous derivative, is the derivative necessarily computable? The answer turns out to be no. This raises the question: are there regularity conditions which we can impose on the initial computable function (e.g. being C^n for some fixed n, or C^∞) which insure the computability of the derivative. Here the answer is yes. The problem is to find the right conditions.

Theorem 1, essentially due to Myhill [1971], asserts that there exists a computable C^1 function whose derivative is not computable. Theorem 2, due to the authors [1983a], asserts that, on the other hand, if the initial function is C^2, then the derivative is computable. Thus the cutoff point must lie somewhere between C^1 and C^2. Where is it located? We give a slight strengthening of Myhill's example, which shows that the function f can be twice differentiable (but not continuously so) and still give a noncomputable f'. Hence the cutoff is pinned between "twice differentiable" (Theorem 1) and "twice continuously differentiable" (Theorem 2).

A curious result which emerges from Theorems 1 and 2 is the following: There exists a function on [0, 1] which is effectively continuous at each point of [0, 1] but not effectively uniformly continuous on [0, 1] (Corollary 2b). Thus the classical theorem that a continuous function on a compact set is uniformly continuous does not effectivize.

An immediate consequence of Theorem 2 (Corollary 2a) asserts that if a computable function f is C^∞, then each derivative $f^{(n)}$ is computable. This raises the question: Is the sequence of derivatives $\{f^{(n)}\}$ computable, effectively in n? The answer is no, as we show in Theorem 3 (cf. Pour-El, Richards [1983a]).

We turn now to Section 2, on analytic functions. Let f be a computable function which is analytic (but with no computability assumptions on its derivatives). Here, in contrast to the C^∞ case, everything effectivizes—at least for compact domains. In particular, the sequence of derivatives $\{f^{(n)}\}$ and the sequence of Taylor coefficients are computable. Surprisingly enough, even analytic continuation is computable. These results are spelled out in Proposition 1.

In view of the rather pervasive effectiveness cited above, the following result seems a bit surprising. There exists an entire function which is computable on every compact disk, but not computable over the whole complex plane (Theorem 4).

Section 3 presents the Effective Modulus Lemma (Theorem 5), a technical result which is useful for the creation of counterexamples. We use it to produce a continuous function f which is sequentially computable—i.e. f maps every computable sequence $\{x_n\}$ onto a computable sequence $\{f(x_n)\}$—but such that f is not computable (Theorem 6). A second application, involving the wave equation, is cited without proof (cf. Pour-El, Richards [1981]).

The chapter closes with Section 4, translation invariant operators. The section begins with a detailed account of translation invariance. Many of the standard operators of analysis and physics are translation invariant. We prove a theorem about these operators (Theorem 7) which has several applications. In particular, Theorem 2 from Section 1 is an immediate corollary of Theorem 7. So too are the extensions of Theorem 2 to partial derivatives. A deeper application of Theorem 7 involves weak solutions of the wave equation (Theorem 8).

1. C^n Functions, $1 \leq n \leq \infty$

Here we present the key results about the computability theory of derivatives, as outlined in the introduction to this chapter.

Note. We recall that a function f, of one or several variables, is said to be C^n if all derivatives of order $\leq n$ exist and are continuous. We say that f is C^∞ if f is C^n for all n.

As noted above, our first theorem is essentially due to Myhill [1971].

Theorem 1 (Noncomputability of the derivative for a computable C^1 function). *There exists a computable function f on $[0, 1]$ such that f is C^1, but the derivative f' is not computable. Furthermore, the function f can be chosen to be twice differentiable (but not twice continuously differentiable).*

Proof. We begin by describing the derivative $f'(x)$, and then obtain $f(x)$ by integration:

$$f(x) = \int_0^x f'(u)\, du.$$

The derivative $f'(x)$ will be a superposition of countably many "pulses", and we start by taking a canonical C^∞ pulse function $\varphi(x)$:

$$\varphi(x) = \begin{cases} e^{-x^2/(1-x^2)} & \text{for } |x| < 1 \\ 0 & \text{for } |x| \geq 1. \end{cases}$$

Then $\varphi \in C^\infty$, φ has support on $[-1, 1]$, and $\varphi(0) = 1$. The sequence of derivatives $\{\varphi^{(n)}(x)\}$ is computable, effectively in n.

Let $a: \mathbb{N} \to \mathbb{N}$ be a one-to-one recursive function listing a recursively enumerable nonrecursive set A. We can assume that $0 \notin A$. Define the n-th pulse $\varphi_n(x)$ by:

$$\varphi_n(x) = \varphi[2^{(n+a(n)+2)} \cdot (x - 2^{-a(n)})].$$

Then $\varphi_n(x)$ is a pulse of height 1 and half width $2^{-(n+a(n)+2)}$, centered on the point $x = 2^{-a(n)}$.

We now give a (noneffective) description of the derivative $f'(x)$:

$$f'(x) = \sum_{k=0}^{\infty} 4^{-a(k)} \varphi_k(x).$$

We observe that the sequence of partial sums

$$f'_n(x) = \sum_{k=0}^{n} 4^{-a(k)} \varphi_k(x)$$

is computable, effectively in n, since the sequences $\{4^{-a(n)}\}$ and $\{\varphi_n(x)\}$ are both computable. Moreover, $f'_n(x) \to f'(x)$ uniformly (although not necessarily effectively) as $n \to \infty$, since $a(n)$ gives a one-to-one listing of the set A, and hence the series converges uniformly by comparison with the series $\sum 4^{-a}$. Thus the limit function $f'(x)$ is continuous.

We assert that $f'(x)$ is also differentiable (although its derivative is discontinuous). For the individual pluses φ_n are C^∞ and their supports approach the point $x = 0$. Hence $f'(x)$ is differentiable except perhaps at $x = 0$.

We now show that near $x = 0$ the graph of $f'(x)$ is bounded between the two parabolas $y = \pm 4x^2$. As a first step, we observe that the pulses $4^{-a(n)} \varphi_n(x)$, centered at the points $x = 2^{-a(n)}$, have heights $y = 4^{-a(n)} = x^2$ when $x = 2^{-a(n)}$. However, the pulses have finite width, and we cannot just consider the central point $x = 2^{-a(n)}$. But the half-width of the pulse is $\leq (1/2) \cdot 2^{-a(n)}$ (with room to spare), and so for *any* point x in support(φ_n), $x \geq (1/2) \cdot 2^{-a(n)}$. Thus, for *any* such x, $2x \geq 2^{-a(n)}$, and $4x^2 \geq 4^{-a(n)}$ = amplitude of n-th pulse.

Since $f'(x)$ is bounded between the two parabolas $y = \pm 4x^2$, $f''(0)$ exists and is zero. Hence $f'(x)$ is differentiable at all points.

To complete the proof, we must show that $f'(x)$ is not computable, but that its antiderivative $f(x)$ is computable. We begin with $f'(x)$. As a first step, we observe that the pulses φ_n in the series for f' have disjoint supports. For the half-width of

φ_n is $2^{-(n+a(n)+2)} \leq 2^{-a(n)-2}$, whereas the pulse-center $x = 2^{-a(n)}$ differs from its nearest possible neighbor $x = 2^{-a(n)-1}$ by a distance of $2^{-a(n)-1}$.

Recall that each pulse φ_n, centered at $x = 2^{-a(n)}$, has height 1 and is multiplied by $4^{-a(n)}$. Hence the value of $f'(x)$ at the point $x = 2^{-a}$ is given by:

$$f'(2^{-a}) = \begin{cases} 4^{-a} & \text{if } a = a(n) \text{ for some } n, \\ 0 & \text{otherwise.} \end{cases}$$

That is, $f'(2^{-a}) = 4^{-a}$ if $a \in A$, and $f'(2^{-a}) = 0$ if $a \notin A$.

The sequence of values $\{f'(2^{-a})\}$, $a = 0, 1, 2, \ldots$, is not a computable sequence of reals. For suppose it were. Then there would exist a computable double sequence of *rationals* $\{r_{ak}\}$ such that $|r_{ak} - f'(2^{-a})| \leq 2^{-k}$ for all k and a (Definition 5a, Section 2, Chapter 0). From this we could derive the computable rational sequence $\{r'_a\}$ given by $r'_a = r_{a,2a+2}$, and we would have $|r'_a - f'(2^{-a})| \leq (1/4) \cdot 4^{-a}$ for all a. Since $\{r'_a\}$ is a computable rational sequence, exact comparisons involving $\{r'_a\}$ can be made effectively. This would give a decision procedure for the set A: $a \in A$ if $r'_a \geq (1/2)4^{-a}$, and $a \notin A$ otherwise. Since A is not recursive, this is a contradiction.

Now we recall (Definition A in Chapter 0) that a computable function must be sequentially computable. Since $\{2^{-a}\}$ is a computable sequence of reals which is mapped by f' onto a noncomputable sequence $\{f'(2^{-a})\}$, the function f' is not computable.

To show that the antiderivative $f(x)$ is computable, we reason as follows. Since the n-th pulse $4^{-a(n)}\varphi_n(x)$ has height $\leq 4^{-a(n)}$, and since this pulse has half-width $2^{-(n+a(n)+2)}$, its antiderivative

$$\Phi_n(x) = \int_0^x 4^{-a(n)}\varphi_n(u)\, du \leq (4^{-a(n)})(2 \cdot 2^{-(n+a(n)+2)}) \leq 2^{-n}.$$

Hence the series for $f(x)$,

$$f(x) = \sum_{k=0}^{\infty} \Phi_k(x),$$

converges uniformly and effectively by comparison with $\sum 2^{-k}$. Furthermore the sequence of summands $\{\Phi_k(x)\}$ is computable. Hence by Theorem 4 in Chapter 0, f is computable. This proves Theorem 1. □

In the opposite direction, we have:

Theorem 2 (Computability of the derivative for computable C^2 functions). *Let $[a, b]$ be an interval with computable endpoints. Let f be a computable function on $[a, b]$ which is C^2. Then the derivative f' is computable. Moreover, the hypothesis that f is C^2 could be replaced by the weaker hypothesis that f' is effectively uniformly continuous.*

Before proving Theorem 2 we give two corollaries.

Corollary 2a (Computability of each derivative of a computable C^∞ function). *If f is computable on $[a, b]$ and f is C^∞—i.e. infinitely differentiable—then the n-th derivative $f^{(n)}$ is computable for each fixed n.*

Proof. This is an immediate consequence of Theorem 2. □

For the next corollary, we need a definition of "pointwise effective continuity". As expected, a function g is *effectively continuous at a point x* if there is a recursive function $d: \mathbb{N} \to \mathbb{N}$ such that $|y - x| \leq 1/d(N)$ implies $|g(y) - g(x)| \leq 2^{-N}$. A function which is effectively continuous at each point of its domain will be called, for simplicity, "effectively continuous".

Corollary 2b (Effective continuity does not imply effective uniform continuity). *There exists a function g on $[0, 1]$ which is effectively continuous at each point $x \in [0, 1]$, but which is not effectively uniformly continuous.*

Proof. We combine the results of Theorems 1 and 2. By Theorem 1, there exists a computable function f on $[0, 1]$ whose derivative f' is itself differentiable but not computable. Let $g = f'$.

By Theorem 2, g is not effectively uniformly continuous. For this theorem asserts that if $g = f'$ were effectively uniformly continuous, then it would be computable, which it is not.

On the other hand, since g is differentiable, it is effectively continuous at each point x. To see this: Fix x. Let $c = g'(x)$. Then $\lim_{y \to x} [(g(y) - g(x))/(y - x)] = c$. Let M be an integer with $M \geq |c| + 1$. Then for all points y sufficiently close to x, $|g(y) - g(x)| \leq M|y - x|$. (The points y which are not close to x are irrelevant.) Thus to achieve $|g(y) - g(x)| \leq 2^{-N}$, it suffices to take $|y - x| \leq 1/M \cdot 2^N$, so that an effective modulus of continuity is given by $d(N) = M \cdot 2^N$. □

Proof of Theorem 2. First we verify that if f is C^2, then f' is effectively uniformly continuous. For if f'' is continuous, then there is some integer M with $|f''(x)| \leq M$ for all $x \in [a, b]$. Then, by the Mean Value Theorem, f' satisfies

$$|f'(x) - f'(y)| \leq M \cdot |x - y|$$

for all $x, y \in [a, b]$. Hence f' is effectively uniformly continuous.

From here on, we will merely assume that f' is effectively uniformly continuous. This means (cf. Definition A in Chapter 0) that there is a recursive function $d(N)$ such that, for all $x, y \in [a, b]$, $|x - y| \leq 1/d(N)$ implies $|f'(x) - f'(y)| \leq 2^{-N}$. We can assume that $d(N)$ is a strictly increasing function.

The assertion (∗) below is a technicality, but it must be dealt with. The problem, of course, is to keep y_{kN} within the interval $[a, b]$, since we cannot decide effectively which of the boundary points a, b is closer to x_k.

(∗) *Let $\{x_k\}$ be a computable sequence of real numbers in $[a, b]$. Then there exists an integer N_0 and a computable double sequence $\{y_{kN}\}$, $y_{kN} \in [a, b]$, such that for each k*

1. C^n Functions, $1 \leqslant n \leqslant \infty$

and $N \geqslant N_0$,

$$\text{either} \quad y_{kN} = x_k + \frac{1}{d(N)} \quad \text{or} \quad y_{kN} = x_k - \frac{1}{d(N)}.$$

Proof of (∗). First we find suitable rational approximations to the computable reals a and b, and to the computable real sequence $\{x_k\}$. For a and b, we merely observe that there exist rational numbers A and B with $|a - A| \leqslant (b - a)/1000$ and $|b - B| \leqslant (b - a)/1000$. For $\{x_k\}$, we readily deduce from Definition 5a, Section 2, Chapter 0, that there exists a computable sequence of rationals $\{X_k\}$ with $|X_k - x_k| \leqslant (b - a)/1000$ for all k. We also choose N_0 so that $N \geqslant N_0$ implies $1/d(N) \leqslant (b - a)/1000$. Now, working with the computable *rational* sequence $\{X_k\}$, we set $y_{kN} = x_k + (1/d(N))$ if X_k is closer to A than to B, and set $y_{kN} = x_k - (1/d(N))$ otherwise. This is an effective procedure which satisfies all of the conditions set down in (∗). □

Now we come to the body of the proof. We want to show that f' is computable, and by Definition A in Chapter 0, this means that f' is (i) sequentially computable and (ii) effectively uniformly continuous. We have (ii) by hypothesis. Now, as we recall, (i) means that if $\{x_k\}$ is a computable sequence, then $\{f'(x_k)\}$ is computable. To compute $\{f'(x_k)\}$, we proceed as follows:

Take $\{y_{kN}\}$ as in (∗) above. Recall that the points y_{kN} are approximations to the x_k, lying either above or below x_k and spaced an exact distance $1/d(N)$ away from x_k. Now consider the difference quotients

$$D_{kN} = \frac{f(y_{kN}) - f(x_k)}{y_{kN} - x_k}.$$

Since the function f itself is computable, $\{D_{kN}\}$ is computable, effectively in k and N. Now by the mean value theorem,

$$D_{kN} = f'(\xi)$$

for some $\xi = \xi_{kN}$ between x_k and y_{kN}. (It makes no difference here whether we can compute ξ effectively or not. It suffices that ξ exists.) By definition of y_{kN}, the distance $|y_{kN} - x_k| = 1/d(N)$, and hence $|\xi - x_k| < 1/d(N)$. Finally, by definition of the modulus of continuity $d(N)$, $|\xi - x_k| < 1/d(N)$ implies $|f'(\xi) - f'(x_k)| \leqslant 1/2^N$. Hence

$$|D_{kN} - f'(x_k)| = |f'(\xi) - f'(x_k)| \leqslant 1/2^N.$$

Thus the computable sequence $\{D_{kN}\}$ converges to $\{f'(x_k)\}$ as $N \to \infty$, effectively in N and k. Hence $\{f'(x_k)\}$ is computable, as desired. This proves Theorem 2. □

Theorem 3 (Noncomputability of the sequence of n-th derivatives). *There exists a computable function f on $[-1, 1]$ which is C^∞, but for which the sequence of derivatives*

$\{f^{(n)}(x)\}$ is not computable. In addition, f can be chosen so that the sequence of values $\{|f^{(n)}(0)|\}$ is not bounded by any recursive function of n.

Proof. Take a one-to-one recursive function $a: \mathbb{N} \to \mathbb{N}$ which generates a recursively enumerable nonrecursive set A. We assume that $0, 1 \notin A$. Let $w(n)$ be the "waiting time"

$$w(n) = \max\{m: a(m) \leq n\}$$

defined in Chapter 0, Section 1. As we saw in Chapter 0, $w(n)$ is not bounded above by any recursive function. (Otherwise, as we recall, there would be an effective procedure for deciding whether or not an arbitrary integer belongs to A.)

We shall construct a computable C^∞ function f such that $|f^{(n)}(0)| \geq w(n)$ whenever $w(n) \geq n$. The exceptional cases where $w(n) < n$ are of no interest, since we know that $w(n)$ is not bounded above by any recursive function, and clearly this failure to be recursively bounded cannot involve those values n for which $w(n) < n$.

We define R_n recursively by setting $R_0 = 1$,

$$R_n = 2nR_{n-1}^n.$$

(Although we do not need to know the exact size of R_n, we observe that $2^{n!} \leq R_n \leq 2^{n^n}$ for $n \geq 1$.)

We now construct the desired function f. Let

$$f(x) = \sum_{k=0}^{\infty} R_k^{1-a(k)} \cdot \cos\left(R_k x + \frac{\pi}{4}\right), \qquad x \in [-1, 1].$$

By Theorem 4 of Chapter 0, f is computable, since the terms of the series are computable (effectively in k), and the series is effectively uniformly convergent (being dominated by R_k^{-1} since $0, 1 \notin A$ and thus $a(k) \geq 2$).

For fixed n, the series for the n-th derivative is also effectively uniformly convergent: since the series is

$$f^{(n)}(x) = \sum_{k=0}^{\infty} R_k^{n+1-a(k)} \begin{Bmatrix} \pm\cos \\ \pm\sin \end{Bmatrix} \left(R_k x + \frac{\pi}{4}\right),$$

and $a(k) \leq n+1$ for only finitely many k. Thus, again by Theorem 4 of Chapter 0, $f^{(n)}(x)$ exists and is a computable function. In particular, since $f^{(n)}$ exists for all n, f is C^∞.

We now show that $|f^{(n)}(0)| \geq w(n)$ for $w(n) \geq n \geq 3$. Take such an n, and for convenience set

$$m = w(n).$$

In the series for $f^{(n)}(x)$, the m-th term (which turns out to dominate all of the others) is

$$R_m^{n+1-a(m)} \cdot \{\pm\sin \text{ or } \cos(\ )\}.$$

1. C^n Functions, $1 \leq n \leq \infty$

Hence $|(m\text{-th term})(x = 0)| = R_m^{n+1-a(m)}\sqrt{1/2}$, since the sin or cos is evaluated at $\pi/4$. Since $a(m) \leq n$ (by definition $m = w(n)$ is the last k for which $a(k) \leq n$), we obtain

$$|(m\text{-th term})(x = 0)| \geq R_m\sqrt{1/2}.$$

We must show that the sum of all other terms in the series for $f^{(n)}(x)$ is smaller than the m-th term. Consider first the previous terms, involving $k < m$. These terms

$$R_k^{n+1-a(k)} \cdot \{\pm\sin \text{ or } \cos(\ )\}, \qquad k < m,$$

are dominated by R_k^n (since $a(k) \geq 2$), and $R_k^n \leq R_k^m$ (here we use the fact that $m = w(n) \geq n$). Furthermore, there are m such terms. Since $R_m = 2mR_{m-1}^m$, the sum of these previous terms (i.e. the terms with $k < m$) has absolute value at most $R_m/2$. Since $\sqrt{1/2} - (1/2) > 1/10$,

$$|m\text{-th term}| - \left|\sum_{k=0}^{m-1} (k\text{-th term})\right| > R_m/10.$$

Now consider the terms with $k > m$. These give $a(k) \geq n + 1$ (again since $m = w(n)$ is the last k for which $a(k) \leq n$). For at most one value of k is $a(k) = n + 1$. Otherwise $a(k) > n + 1$, and the exponent $n + 1 - a(k)$ in

$$R_k^{n+1-a(k)} \cdot \{\pm\sin \text{ or } \cos(\ )\}$$

is negative. We bound the possible term with $a(k) = n + 1$ by $R_k^0 = 1$, and the sum of the other terms (with $k > m$) by $\sum_{k=1}^{\infty} R_k^{-1} \leq 1$ (since $R_k \geq 2^k$). Thus the effect of all terms with $k > m$ is dominated by $1 + 1 = 2$. Hence for $f^{(n)}(0)$, which is the sum of all the terms (with $k = m$, $k < m$, and $k > m$):

$$|f^{(n)}(0)| \geq (R_m/10) - 2.$$

Now for $m = w(n) \geq n \geq 3$,

$$(R_m/10) - 2 \geq m.$$

(In fact, since $R_m \geq 2^{m!}$, this inequality is absurdly weak.) Combining the last two displayed inequalities, we have, for $w(n) \geq n \geq 3$:

$$|f^{(n)}(0)| \geq m = w(n),$$

as desired. This proves Theorem 3. □

The final remarks in this section all relate to Theorem 2. As we recall, this theorem gives conditions under which the derivative *is* computable (i.e. if f is computable and C^2, then f' is computable).

Partial derivatives. Theorem 2 extends in an obvious way to partial derivatives. Thus if $f(x, y)$ is computable and C^2 on a computable rectangle in $\mathbb{R}^2$, then $\partial f/\partial x$ and $\partial f/\partial y$ are computable. If $f(x, y)$ is computable and C^∞ on this rectangle, then each partial derivative

$$\frac{\partial^{m+n} f}{\partial x^m \, \partial y^n}$$

is computable. Similar results hold for $\mathbb{R}^q$, $q > 2$.

In Section 4 we give a general result (Theorem 7) from which Theorem 2, its extension to partial derivatives, and numerous other results are immediate corollaries.

It is natural to ask whether Theorem 2 extends (A) to noncompact domains, or (B) to sequences of functions. The answer is "no" in both cases, as the following two remarks show.

Remark A (The noncompact case). Theorem 2 breaks down if the domain $[a, b]$ is replaced by $\mathbb{R}$. For there exists a C^∞ function f on $\mathbb{R}$ which is computable on $\mathbb{R}$ (in the sense of Definition A'' of Chapter 0), but such that $f'(x)$ is not computable on $\mathbb{R}$.

Here is the construction of f. Start with the C^∞ function φ used in the proof of Theorem 1 above. Let $\psi(x) = \varphi[x - (1/2)]$, so that $\psi'(0) = c > 0$. Set $\psi_k(x) = (1/k) \cdot \psi(k^2 x)$, so that ψ_k has amplitude $1/k$ but ψ_k' has amplitude $= \text{Const} \cdot k$. As previously, let $a \colon \mathbb{N} \to \mathbb{N}$ be a recursive function generating a recursively enumerable nonrecursive set A in a one-to-one manner. Let

$$f(x) = \sum_{k=2}^{\infty} \psi_k(x - a(k)).$$

Then it is easy to verify that f is C^∞ and computable on $\mathbb{R}$.

We now show that $f'(x)$ is not computable. This is done by showing that the sequence $\{f'(n)\}$ is not computable. Let $k \geq 2$. Then

$$f'(n) = \begin{cases} k\varphi'(-1/2) & \text{if } n = a(k), \\ 0 & \text{otherwise.} \end{cases}$$

Suppose $\{f'(n)\}$ were computable. Then, since $\varphi'(-1/2)$ is computable, the sequence $\{\mu_n\}$ where

$$\mu_n = \frac{f'(n)}{\varphi'(-1/2)}$$

would also be computable. Hence A would be recursive. For $n \in A$ if and only if one of the following holds: $n = a(0)$ or $n = a(1)$ or $\mu_n \geq 2$. Thus we have a contradiction. □

It is interesting to note that the sequence $\{f'(n)\}$ grows faster than any recursive function. For, except for a finite number of values,

$$\max_{p \leqslant n} f'(p) = \max_{a(k) \leqslant n} (k\varphi'(-1/2)) = \varphi'(-1/2)w(n),$$

where $w(n) = \max\{k: a(k) \leqslant n\}$ is the function defined in the Waiting Lemma of Chapter 0, Section 1. As we showed there, $w(n)$ grows faster than any recursive function of n.

Remark B (Sequences of functions). Theorem 2 also breaks down for sequences of functions. For there exists a computable sequence $\{f_k\}$ on $[0, 1]$, such that each f_k is C^∞, but the sequence of derivatives $\{f'_k\}$ is not computable.

To construct this example, simply take the function f from Remark A and set

$$f_k(x) = f(x + k), \quad 0 \leqslant x \leqslant 1. \quad \square$$

A final note. The explanation of the pathological behavior exhibited in Remarks A and B above is that, in both cases, there is a noncompact domain. In Remark A the domain is $\mathbb{R}$; whereas in Remark B the sequence $\{f_k\}$ can be viewed as a function on the noncompact set $[0, 1] \times \mathbb{N}$.

We will see further instances of these phenomena in the section which follows.

2. Analytic Functions

Here we deal with the computability theory of analytic functions of a complex variable. In what follows, the function f will be assumed to be analytic on some region Ω in the complex plane, and we will also assume that f is computable in the sense of Chapter 0 (details to follow). However, we do *not* assume that f is "computably analytic"—i.e. we do not assume that the derivatives or the power series expansion of f are computable. In fact, this assumption is redundant, as we show in the proposition below.

First some technicalities. In Chapter 0 we defined "computability" for a function f whose domain was a computable closed rectangle in $\mathbb{R}^q$—i.e. a rectangle with computable coordinates for its corners. To do complex analysis, one needs more general domains. For, in complex analysis, we are usually given a function f which is analytic on a connected open region Ω. The shape of Ω can be quite complicated, and frequently f will have no analytic extension beyond Ω. We want to describe "computability" on arbitrary compact subsets K of Ω. The natural approach, following Chapter 0, is to begin by defining computability on a closed region Δ which is a finite union of computable rectangles. Then any compact set $K \subseteq \Omega$ can be covered by such a region Δ, $K \subseteq \Delta \subseteq \Omega$. Thus:

Definition. (a) Let $\Delta \subseteq \mathbb{C}$ be a finite union of computable rectangles. A function $f: \Delta \to \mathbb{C}$ is called *computable* if f is computable on each of the constituent computable rectangles in Δ.

(b) A function f defined on a compact set $K \subseteq \mathbb{C}$ is called *computable* if f has a computable extension to a finite union of computable rectangles $\Delta \supseteq K$.

In the proposition below, we deal with all of the standard themes of elementary complex analysis, including analytic continuation.

Proposition 1 (Basic facts for analytic functions). *Let Ω be an open region in $\mathbb{C}$. Let $f: \Omega \to \mathbb{C}$ be analytic in Ω. Suppose f is computable on some computable rectangle D with nonempty interior, $D \subseteq \Omega$. Then:*

(a) *Effective analytic continuation—the function f, originally assumed to be computable only on D, is computable on any compact subset $K \subseteq \Omega$.*

(b) *The sequence of derivatives $\{f^{(n)}(z)\}$ is computable on K, effectively in n.*

(c) *For any computable point $a \in \Omega$, the sequence of Taylor coefficients of $f(z)$ about $z = a$ is computable.*

(d) *For any computable point $a \in \Omega$ and all computable reals $M \geq 0$, the Taylor series for f converges effectively and uniformly in any closed disk $\{|z - a| \leq M\} \subseteq \Omega$.*

Note. This result contrasts with Theorem 3 (for C^∞ functions) in the preceding section. In Theorem 3, each derivative $f^{(n)}(x)$ was computable, but the sequence of derivatives was not. Here the entire sequence of derivatives is computable. Of course, this corresponds to well known distinctions between the properties of C^∞ functions and analytic functions.

We observe that part (a), about the computability of analytic continuation, applies only when the continuation is to a compact region K. For continuations to the entire complex plane, the result fails, as Theorem 4 below shows.

Proof. Since any compact set $K \subseteq \Omega$ can be covered by a finite union of computable rectangles $\Delta \subseteq \Omega$, it suffices to prove the proposition for Δ. Thus we replace the arbitrary compact set K by a set of the special form Δ.

The proof proceeds in several stages. Initially we know only that f is computable on the rectangle D. We first prove parts (b) and (c) for a smaller rectangle D' contained in the interior of D. Then, using this information, we prove parts (d) and (a) in general. Then we come back and prove parts (b) and (c) for the larger region Δ, $\Delta \supseteq K$.

Let γ be the boundary of the rectangle D, where the curve γ is taken in the positive sense. Let D' be any computable rectangle in the interior of D (and hence inside of γ). As already noted, we first prove (b) and (c) of the proposition for the special case of the rectangle D'.

For part (b) on D'. By the Cauchy integral formula, for points $z \in D'$,

$$f^{(n)}(z) = \frac{n!}{2\pi i} \int_\gamma \frac{f(w)}{(w-z)^{n+1}} dw.$$

2. Analytic Functions

Now γ (= boundary of D) consists of a finite number of vertical or horizontal line segments with computable coordinates for their endpoints. Hence, by Corollary 6b in Chapter 0, the integration can be carried out effectively, and $\{f^{(n)}(z)\}$ is computable on D', effectively in n.

For part (c) on D'. It follows immediately that, for any computable point $a \in D'$, the sequence of Taylor coefficients $\{f^{(n)}(a)/n!\}$ is computable.

For part (d) in general. Here assume that we have *any* computable point $a \in \Omega$ for which it is known that the sequence of Taylor coefficients $\{f^{(n)}(a)/n!\}$ is computable.

We wish to show that the Taylor series for f converges effectively and uniformly in any disk $\{|z - a| \leqslant M\} \subseteq \Omega$. Take a larger closed disk $\{|z - a| \leqslant M + \varepsilon\}$, still contained inside the region Ω. Now the standard estimates found in any complex analysis text (e.g. Ahlfors [1953]) are already effective. Namely, let

$$f(z) = f(a) + \cdots + \frac{f^{(n)}(a)}{n!}(z - a)^n + R_{n+1}(z),$$

where

$$R_{n+1}(z) = \frac{1}{2\pi i} \int_{|w-a|=M+\varepsilon} \frac{f(w)(z-a)^{n+1}}{(w-a)^{n+1}(w-z)} \, dw.$$

Then for $|z - a| \leqslant M$, the "error" $R_{n+1}(z)$ satisfies

$$|R_{n+1}(z)| \leqslant \text{Const}\, \frac{M}{\varepsilon}\left(\frac{M}{M+\varepsilon}\right)^n,$$

where

$$\text{Const} = \max\{|f(w)|: |w - a| \leqslant M + \varepsilon\}.$$

Thus the "Const" depends on f, M, and ε, but it does not depend on z or n. This gives effective uniform convergence of $R_{n+1}(z)$ to zero as $n \to \infty$, and hence proves part (d).

Now we turn to part (a)—effective analytic continuation. Take a computable point $a \in D'$ and a closed disk $\{|z - a| \leqslant M\}$ as above. The Taylor series for f about $z = a$ is computable by part (c), and this series is effectively uniformly convergent on the disk $\{|z - a| \leqslant M\}$ by part (d). Hence, by Theorem 4 in Chapter 0, f has a computable analytic continuation to this disk. Now, starting with D' and using a finite chain of overlapping closed disks, we can cover any compact region $\Delta \subseteq \Omega$. Hence by part (d), already proved, f has a computable analytic continuation to Δ. This proves part (a).

Now, as promised, we extend parts (b) and (c) from the small rectangle D' to any finite union of computable rectangles $\Delta \subseteq \Omega$. Let $\Delta^* \subseteq \Omega$ be another finite union of computable rectangles, such that each rectangle in Δ lies in the *interior* of some rectangle in Δ^*. By part (a), already proved, f has a computable analytic continuation to Δ^*. But then our previous proofs of (b) and (c) (for the rectangle D' inside of D) extend mutatis mutandis to all of the rectangles which make up Δ. This completes the proof of Proposition 1. □

We have seen that on compact domains, everything goes as one would expect. However, for noncompact domains, the results are a little more startling. The following theorem is due to Caldwell and Pour-El [1975].

Theorem 4 (An entire function which is computable on every compact disk but not computable over the whole plane). *There exists an entire function f which is computable on any compact domain in $\mathbb{C}$, but such that f is not computable over the whole complex plane. Furthermore, f can be chosen so that, as $x \to \infty$ along the positive real axis, $f(x)$ grows more rapidly than any computable function of x.*

Note. In fact we will have that the sequence of values $f(0), f(1), f(2), \ldots$ is not bounded by any recursive function.

Proof. Let $a: \mathbb{N} \to \mathbb{N}$ be a one-to-one recursive function enumerating a recursively enumerable non recursive set A. We can assume that $0 \notin A$. Now we define f by

$$f(z) = \sum_{m=0}^{\infty} \frac{z^m}{a(m)^m}.$$

Clearly the sequence of Taylor coefficients $\{1/a(m)^m\}$ is computable. Furthermore, the series is effectively uniformly convergent on any compact disk $\{|z| \leq M\}$, where M is an integer >0. To see this, we observe that there are only finitely many values of m with $a(m) \leq M$. For all other m, $a(m) \geq M + 1$, and since $|z| \leq M$, the series is dominated by

$$\sum \left(\frac{M}{M+1}\right)^m.$$

Of course, this dominating series is effectively convergent.

Thus on the disk $\{|z| \leq M\}$, f is the effective uniform limit of a computable sequence of functions (the partial sums of its Taylor series), and so by Theorem 4 of Chapter 0, f is computable.

[It should be noted that the above construction is not effective in M. For, as $M \to \infty$, there is no effective way of telling how many m's there are with $a(m) \leq M$.]

We now show that the sequence of values $f(0), f(1), f(2), \ldots$ is not bounded above by any recursive function. Suppose otherwise, that there is a recursive function g with

$$f(n) \leq g(n) \qquad \text{for all } n.$$

2. Analytic Functions

Now we recall the "waiting time"

$$w(n) = \max \{m: a(m) \leq n\}$$

defined in Section 1 of Chapter 0. As we showed in Chapter 0, $w(n)$ is not bounded above by any recursive function. Now we will show that $w(n) \leq f(2n) \leq g(2n)$, and thus derive a contradiction.

To show that $w(n) \leq f(2n)$. We observe that all of the coefficients in the Taylor series for f are positive, and hence $f(2n) >$ any single term in its Taylor series expansion. We use the term with $m = w(n)$. Then, since by definition $a(m) \leq n$,

$$f(2n) > \left(\frac{2n}{a(m)}\right)^m \geq \left(\frac{2n}{n}\right)^m = 2^m = 2^{w(n)} > w(n).$$

Hence $w(n) < f(2n) \leq g(2n)$, giving a recursive upper bound $g(2n)$ for $w(n)$, and thus giving the desired contradiction. This proves Theorem 4. □

We conclude this section with another counterexample. In Proposition 1 we proved the effectiveness of analytic continuation to compact domains. (Theorem 4 shows that this breaks down for noncompact domains.) Even for compact domains, however, the result fails if we consider *sequences* of analytic functions. The failure for sequences is connected with the fact that analytic continuation is not "well posed" in the sense of Hadamard. That is, analytic functions which are "small" on a compact domain D_1 can grow arbitrarily rapidly when continued to a larger compact domain D_2. This the following example shows.

Example (Failure of effective analytic continuation for sequences of functions). There exists a sequence $\{f_k\}$ of entire functions which is computable on the disk $D_1 = \{|z| \leq 1\}$ but not computable on $D_2 = \{|z| \leq 2\}$.

Since this is similar to several of our other examples, we shall be very terse. Let $a: \mathbb{N} \to \mathbb{N}$ give a one-to-one recursive listing of a recursively enumerable non recursive set A. Delete the value $a(0)$ from A. Define

$$f_{kn}(z) = \begin{cases} \dfrac{z^m}{m} & \text{if } k = a(m) \text{ for some } m, 1 \leq m \leq n, \\ 0 & \text{otherwise.} \end{cases}$$

Clearly $\{f_{kn}\}$ is computable. Let

$$f_k(z) = \begin{cases} \dfrac{z^m}{m} & \text{if } k = a(m) \text{ for some } m \geq 1, \\ 0 & \text{otherwise.} \end{cases}$$

As displayed, $\{f_k\}$ is not computable. But on D_1, f_{kn} converges effectively and uniformly to f_k as $n \to \infty$. Namely when $z \in D_1$, $|f_{kn}(z) - f_k(z)| \leq 1/n$. Thus $\{f_k\}$ is computable on D_1.

On the other hand, when we enlarge the domain to D_2, the values $\{f_k(2)\}$ increase faster than any recursive function of k. To see this: We recall that $f_k(z) = z^m/m$ if $k = a(m)$ for some $m \geq 1$, and $f_k(z) = 0$ otherwise. Set $z = 2$. We observe that $2^m/m \geq m$ for all $m \geq 4$. So, with finitely many exceptions,

$$\max_{j \leq k} f_j(2) \geq \max \{m: a(m) \leq k\} = w(k),$$

where $w(k)$ is the function defined in the Waiting Lemma. By that lemma, $w(k)$ and hence $f_j(2)$, $j \leq k$, are not dominated by any recursive function. □

Real-analytic functions. Part (b) of Proposition 1 above (computability of the sequence of derivatives) extends in a natural way to real analytic functions. For details, see Pour/El, Richards [1983a].

3. The Effective Modulus Lemma and Some of Its Consequences

Our first result, the Effective Modulus Lemma, is useful for the construction of counterexamples. We give this result and then give two applications of it.

Before we state the Effective Modulus Lemma, a brief introduction seems in order. Recall from Chapter 0, Sections 1 and 2, that if $\{r_m\}$ is a computable monotone sequence of reals which converges noneffectively to a limit α, then α is not computable. Examples of this type are used several times in this book. However, sometimes one needs a sharper version of this construction. The Effective Modulus Lemma provides such a sharpening. In this lemma, we have $\{r_m\}$ and α as above (so that, in particular, α is a noncomputable real), but we also have more. Namely, for any computable real sequence $\{\gamma_k\}$, there exists a recursive function $d: \mathbb{N} \to \mathbb{N}$ such that $|\gamma_k - \alpha| \geq 1/d(k)$ for all k. The striking thing here is that, although the moduli $|\gamma_k - \alpha|$ are effectively bounded away from zero, there is no effective way to determine the *signs* of the numbers $(\gamma_k - \alpha)$. Indeed, if there were, then α would be computable—for to compute α, we would merely take for $\{\gamma_k\}$ any recursive enumeration of the rationals.

Both of the applications given below appear to require the full strength of the Effective Modulus Lemma. The first application gives an example of a continuous function f which is sequentially computable—i.e. $\{f(x_n)\}$ is computable whenever $\{x_n\}$ is—but such that f is not computable. This example has an interesting connection with the history of recursive analysis. Originally, Banach and Mazur defined a function f to be "computable" if f was sequentially computable (condition (i) in Definition A of Chapter 0). Later it was realized that this definition was too broad, and condition (ii) of Definition A—effective uniform continuity—was added.

3. The Effective Modulus Lemma and Some of Its Consequences

In Theorem 6 below, we give an example of a *continuous* function f which satisfies the Banach-Mazur condition (i), but which is not "computable" in the modern sense, since it fails to satisfy condition (ii).

Our second application, which we cite without proof (cf. Pour-El, Richards [1981]), shows that all of the above mentioned phenomena can occur for solutions of the wave equation of mathematical physics. Again, the proof of this assertion depends on the Effective Modulus Lemma.

Theorem 5 (Effective Modulus Lemma). *There exists a computable sequence of rational numbers $\{r_m\}$ such that*:

(1) *the sequence $\{r_m\}$ is strictly increasing, $0 < r_m < 1$, and $\{r_m\}$ converges to a noncomputable real number α.*

(2) *the differences $(r_m - r_{m-1})$ do not approach zero effectively.*

(3) *for any computable sequence of reals $\{\gamma_k\}$, there exist recursive functions $d(k)$ and $e(k)$ such that*

$$|\gamma_k - r_m| \geq 1/d(k) \quad \text{for } m \geq e(k).$$

[*In particular, $|\gamma_k - \alpha| \geq 1/d(k)$ for all k.*]

Proof. We begin with a pair of recursively inseparable sets of natural numbers A and B; as we recall, this means that the sets A and B are recursively enumerable and disjoint, and there is no recursive set C with $A \subseteq C$ and $B \cap C = \emptyset$. Let $a: \mathbb{N} \to \mathbb{N}$ and $b: \mathbb{N} \to \mathbb{N}$ be one-to-one recursive functions listing the sets A and B respectively. We assume that $0 \notin A$.

For the time being, we put aside the set B and work with A. We set

$$r_m = \frac{5}{9} + \sum_{n=0}^{m} 10^{-a(n)}, \qquad \alpha = \frac{5}{9} + \sum_{n=0}^{\infty} 10^{-a(n)}.$$

Thus the decimal expansion of α is a sequence of 5's and 6's, with a 6 in the s-th place if and only if $s \in A$. (Since $0 \notin A$, $0 < \alpha < 1$.)

Here we recall briefly some facts which were worked out in detail in Sections 1 and 2 of Chapter 0. The above construction already gives parts (1) and (2) of Theorem 5. Begin with (2): $r_m - r_{m-1} = 10^{-a(m)}$, which cannot approach zero effectively, else we would have an effective test for membership in the set A, and A would be recursive. Now (2) implies (1)—for, since $\alpha - r_m \geq r_{m+1} - r_m$, (2) shows that r_m cannot approach α effectively. We know from Chapter 0 that this forces α to be a noncomputable real.

Now we come to part (3). Take any computable sequence of reals $\{\gamma_k\}$. We must construct the recursive functions $d(k)$ and $e(k)$ promised in (3). To do this, we will use the decimal expansions of the numbers γ_k. However, there is the difficulty that the determination of these expansions may not be effective, because of the ambiguity between decimals ending in 000... and those ending in 999.... To circumvent this difficulty, we use "finite decimal approximations" to the γ_k. These are constructed as follows:

Since $\{\gamma_k\}$ is computable, there exists a computable double sequence of rationals $\{R_{kq}\}$ such that $|R_{kq} - \gamma_k| \leq 10^{-(q+2)}$ (cf. Chapter 0, Section 2). Of course, as we have often noted, exact comparisons between rationals can be made effectively. We define the (q-th decimal for γ_k) to be that decimal of length $q + 1$ which most closely approximates R_{kq}—in case of ties we take the smaller one. Then $|(q$-th decimal for $\gamma_k) - R_{kq}| \leq (1/2)10^{-(q+1)}$ and $|R_{kq} - \gamma_k| \leq (1/10)10^{-(q+1)}$, so that

$$|(q\text{-th decimal for } \gamma_k) - \gamma_k| \leq 10^{-(q+1)}.$$

Now suppose we write out this decimal:

$$(q\text{-th decimal for } \gamma_k) = N_{q0} \cdot N_{q1} N_{q2} \ldots N_{q,q+1}.$$

Then the "digits" $N_{qs} = N_{qs}(k)$ (where $s \leq q + 1$) are recursive in q, s, and k. This is the desired effective sequence of finite decimal approximations to γ_k.

Now we come to the heart of the proof. It is here that we use the other set B in our recursively inseparable pair. We shall give the construction for a particular γ_k, but in a manner which is clearly effective in k. Here is the construction:

List the sets A and B in turn, using the recursive functions a and b. Stop when an integer $s \in A \cup B$ occurs such that either:

a) $s \in A$ and $N_{ss} \neq 6$, $N_{s,s+1} \neq 0$ or 9, or
b) $s \in B$ and $N_{ss} = 6$, $N_{s,s+1} \neq 0$ or 9, or
c) $s \in A \cup B$ and $N_{s,s+1} = 0$ or 9.

This process eventually halts. To prove this, suppose that (c) never occurs. Let C denote the set of integers $\{s: N_{ss} = 6\}$. Since $N_{ss}(k)$ is recursive, C is recursive, effectively in k. Hence, since the sets A and B are recursively inseparable, we cannot have $A \subseteq C$ and $B \cap C = \emptyset$. If $A \not\subseteq C$ then we have (a): for there is some $s \in A$ with $s \notin C$, which means that $s \in A$ and $N_{ss} \neq 6$, and—since we have ruled out (c)—this is precisely what we need for (a). Similarly, if $B \cap C \neq \emptyset$ then we have (b). This covers all cases; for we have shown that if (c) fails, then we eventually arrive at either (a) or (b)—i.e. the process halts. Furthermore, the process can be carried out for all of the γ_k, by using an effective procedure which returns to each k infinitely often.

Now we define the functions $d(k)$ and $e(k)$. Let s be the first occurrence of a value $s = a(n)$ or $s = b(n)$ for which (a), (b), or (c) holds. Then set

$$d(k) = 10^{s+1}, \qquad e(k) = n.$$

We must show that $|\gamma_k - r_m| \geq 1/d(k)$ for $m \geq e(k)$.

First we examine the case where the above process terminates in (c). Consider the true (i.e. exact) decimal expansion for γ_k. By (c), the $(s + 1)$st decimal digit for γ_k is 8, 9, 0, or 1 (allowing for errors in the decimal approximation). By contrast, the $(s + 1)$st digit for r_m is 5 or 6. This gives $|\gamma_k - r_m| \geq 10^{-(s+1)} = 1/d(k)$, as desired.

Now we turn to the case where the process terminates in (a) or (b). This will be a standing assumption throughout the remainder of the proof.

Recall that the s-th decimal digit for α is a 6 if $s \in A$ and a 5 if $s \notin A$. Furthermore, since $a(n)$ gives a one-to-one listing of A, and since $A \cap B = \emptyset$, the s-th decimal digit for both r_m and α is determined as soon as some $a(n)$ or $b(n)$ equals s (the s-th digit is 6 if $s \in A$, 5 if $s \in B$). By definition of $e(k)$, this occurs as soon as $m \geq e(k)$. Thus, for $m \geq e(k)$, the s-th digit for r_m coincides with that for α.

Now suppose that the process terminates in (a). Then $s \in A$, so that the s-th digit for r_m is 6. However, N_{ss} = the s-th digit for γ_k is *not* 6. Finally, $N_{s,s+1} \neq 0$ or 9, which guarantees that N_{ss} is the "true" s-th decimal digit for γ_k (i.e. that N_{ss} is not "off by 1" due to an error in approximation). Hence $|\gamma_k - r_m| \geq 10^{-(s+1)} = 1/d(k)$, as desired.

The case (b) is handled similarly, and this completes the proof of Theorem 5. □

Now as promised above, we give two results which depend on the Effective Modulus Lemma.

Theorem 6 (A continuous function which is sequentially computable but not computable). *There exists a continuous function f on $[0, 1]$ such that f is sequentially computable—i.e. f maps computable sequences $\{x_n\}$ onto computable sequences $\{f(x_n)\}$—but f is not computable. Furthermore, f can be chosen to be differentiable (but not continuously differentiable).*

Proof. We use the results and notation of the Effective Modulus Lemma (Theorem 5 above). Let $\{r_m\}$ be as in the lemma, and set

$$a_m = r_m - r_{m-1},$$
$$b_m = \min(a_m, a_{m-1}), \quad m \geq 2.$$

Since $\{r_m\}$ is strictly increasing, $b_m > 0$. Finally, we use a computable C^∞ pulse function $\varphi(x)$ with support on $[-1, 1]$ as in the proof of Theorem 1 in Section 1. Now we define f by:

$$f(x) = \sum_{m=2}^{\infty} a_m^2 \cdot \varphi[2^m b_m^{-1}(x - r_{m-1})].$$

Since $a_m = r_m - r_{m-1}$ and $\{r_m\}$ converges, $\sum a_m$ converges, and since $0 < a_m < 1$, $\sum a_m^2$ converges (although not effectively). Hence the series for f is uniformly convergent (again not effectively), and f is continuous.

Now the individual pulses in the series for f have

$$\text{amplitudes} = a_m^2 \quad (\text{since } \varphi(0) = 1),$$
$$\text{half-widths} = 2^{-m} b_m \leq 2^{-m},$$

and are centered at the points $x = r_{m-1}$. By definition of b_m (and since $m \geq 2$) these pulses do not overlap. Hence $f(x)$ is C^∞ except at the point $\alpha = \lim r_m$. But $f'(\alpha)$ exists and is zero, since the graph of $f(x)$ is squeezed between the two parabolas $y = \pm 4(x - \alpha)^2$, which are tangent to the x-axis at $x = \alpha$.

To see this: The amplitude of the m-th pulse is a_m^2, and the pulse is centered at $x = r_{m-1}$. However, the pulses have finite width, and we must consider how close the support of the m-th pulse comes to the limit point α. Now the half-width of the m-th pulse is $\leqslant (1/2)b_m \leqslant (1/2)a_m = (1/2)(r_m - r_{m-1}) < (1/2)(\alpha - r_{m-1})$. Thus the support of the m-th pulse reaches less than half of the way from r_{m-1} to α. Hence, for $x \in$ support of m-th pulse, $(\alpha - x) \geqslant (1/2)(\alpha - r_{m-1})$. Thus $4(\alpha - x)^2 \geqslant (\alpha - r_{m-1})^2 \geqslant (r_m - r_{m-1})^2 = a_m^2 =$ amplitude, as desired.

Thus $f'(\alpha) = 0$, and hence f is differentiable at all points.

We now show that f is not effectively uniformly continuous, and hence not computable. Recall that the pulses in the series for f have disjoint supports. The m-th pulse has amplitude $= a_m^2$ and half-width $\leqslant 2^{-m}$. Thus the half-widths approach zero effectively, and effective uniform continuity would force the amplitudes to do likewise. But by (2) in the Effective Modulus Lemma, we have that $a_m = r_m - r_{m-1}$ does not approach zero effectively, and so neither does a_m^2.

Finally, we show that f is sequentially computable, i.e. that $\{f(\gamma_n)\}$ is computable for any computable sequence of reals $\{\gamma_n\}$. It is here that we use part (3) of the Effective Modulus Lemma. We recall that (3) gives recursive functions $d(k)$ and $e(k)$ such that

$$|\gamma_k - r_m| \geqslant 1/d(k) \qquad \text{for } m \geqslant e(k).$$

This allows us to sum the series for $f(\gamma_k)$ in a finite number of steps, effectively in k. Namely let

$$M(k) = \max(d(k), e(k)).$$

Then to compute $f(\gamma_k)$, we simply compute the first $M(k)$ terms in the series for f; all other terms vanish at $x = \gamma_k$. To see this, we recall that the m-th pulse is centered on r_{m-1} and has half-width $\leqslant 2^{-m}$. When $m > M(k)$, then

$$2^{-m} < 2^{-d(k)} < 1/d(k);$$

but $m - 1 \geqslant e(k)$ and hence

$$|\gamma_k - r_{m-1}| \geqslant 1/d(k),$$

so that the support of the m-th pulse does not contain the point $x = \gamma_k$. This proves Theorem 6. □

Another result which depends on the Effective Modulus Lemma is the following. We omit its proof; it can be found in Pour-El, Richards [1981].

Theorem* (The wave equation with computable initial data but a noncomputable solution). *Consider the wave equation*

$$\frac{\partial^2 u}{\partial x^2} + \frac{\partial^2 u}{\partial y^2} + \frac{\partial^2 u}{\partial z^2} - \frac{\partial^2 u}{\partial t^2} = 0,$$

with the initial conditions

$$u(x, y, z, 0) = f(x, y, z),$$

$$\frac{\partial u}{\partial t}(x, y, z, 0) = 0.$$

There exists a computable function $f(x, y, z)$ such that the unique solution $u(x, y, z, t)$ is continuous and sequentially computable on $\mathbb{R}^4$, but the solution $u(x, y, z, 1)$ at time $t = 1$ is not computable on $\mathbb{R}^3$.

It is worth remarking that the main point of Theorem*—the wave equation with computable initial data f can have a noncomputable continuous solution u—will be proved in Chapter 3, Section 5 below. The proof will be based on the First Main Theorem and will be quite short.

4. Translation Invariant Operators

This section is not essential for the rest of the book and could be omitted on first reading. It marks a transition between the concrete topics treated in Part I and the more general notions introduced in Parts II and III. In Part I we have mainly dealt with specific problems, and we have used one notion of computability—that introduced in Chapter 0. In parts II and III we will introduce an infinite class of "computability structures", of which the Chapter 0 notion becomes a special case. Furthermore, rather than treating problems one at a time, we shall seek general theorems which encompass a variety of applications.

This section harks back to Part I in that we are still dealing with the classical— Chapter 0—notion of computability. On the other hand, it reflects the spirit of Parts II and III in that we give a general theorem which has several applications. In order to state this theorem, we need the idea of a translation invariant operator.

We now give a brief introduction to translation invariant operators. In a technical sense, this introduction is unnecessary. Readers who prefer a pure definition/ theorem/proof style of presentation can turn to the conditions (1)–(3) below, and omit the explanations which precede them.

The following is a heuristic discussion, designed to show that the formal conditions which we give below do embody what we mean by "translation invariance". Begin with translation itself. There are two notions of "translation" which are commonly encountered in real analysis. The first is "discrete translation": the translation of a function f through a displacement a, given by the mapping $f(x) \to f(x - a)$. The second (which could be called "continuous translation") is convolution, the mapping from f into $f * g$ given by the convolution formula

$$(f * g)(x) = \int_{-\infty}^{\infty} f(x - a)g(a)\, da.$$

Actually we shall need the q-dimensional analog, in which $x, a \in \mathbb{R}^q$. This is

$$(f * g)(x) = \int \cdots \int_{\mathbb{R}^q} f(x - a)g(a)\, da_1 \ldots da_q.$$

The two versions of "translation"—discrete translation and convolution—are closely related. For it is a small step from the single discrete translation $f(x) \to f(x - a)$ above to a finite linear combination of such translations. Namely, consider a finite sequence $f(x - a_1), \ldots, f(x - a_n)$ of translates of f, and together with these a sequence of coefficients $g_1, \ldots, g_n$. Then the corresponding linear combination is

$$\sum_{i=1}^{n} g_i \cdot f(x - a_i).$$

This is clearly analogous to a Riemann sum for the integral $(f * g)$ above. Since we are only doing heuristics, there is no need to pursue this analogy further. For our purposes it is enough to know that convolution is a natural extension of the idea of a discrete translation. We shall work with convolutions because, in most applications, they are easier to deal with.

Now thinking of convolution as a kind of "continuous translation", we define a *translation invariant operator* T to be one which commutes with convolution, i.e. such that:

$$T(f * g) = (Tf) * g.$$

[To show the analogy with the discrete case, recall that there the operation $f \to f * g$ is replaced by $f(x) \to f(x - a)$. Then, by analogy with the above, translation invariance means that $T[f(x - a)] = (Tf)(x - a)$, just as we would expect.]

Many of the standard operators of analysis and physics are translation invariant: among them the derivative d/dx, the partial derivatives $\partial/\partial x_i$, the Laplace operator, and the solution operators for the wave and heat equations.

For our purposes, we must consider translation invariant operators which satisfy two side-conditions. We now list all of the hypotheses which we will set for the operator T.

Let X and Y be vector spaces of real-valued functions on $\mathbb{R}^q$, and let $T: X \to Y$ be a linear operator such that:
 (1) (Translation invariance.) T commutes with convolution, i.e. $T(f * g) = (Tf) * g$.
 (2) (Compact support). If f has compact support, then so does Tf.
 (3) (Computability for smooth functions.) If $\{\varphi_k\}$ is a sequence of C^∞ functions with compact support on $\mathbb{R}^q$, and if the partial derivatives $\{\varphi_k^{(\alpha)}\}$ (α = multi-index) are computable effectively in k and α, then the sequence $\{T\varphi_k\}$ is computable.

Now we give a general result which holds for operators of this type. By way of preface, we recall the two conditions in the definition of a computable function on

4. Translation Invariant Operators

a compact domain (Definition A, Chapter 0):

(i) sequential computability;
(ii) effective uniform continuity.

These conditions are independent in the sense that, in general, neither implies the other. However, there are situations where (ii) implies (i), and this has numerous applications.

Theorem 7 (Translation invariant operators). *Let T be a linear operator which satisfies* (1)–(3) *above. Let $f: \mathbb{R}^q \to \mathbb{R}^1$ be a computable function with compact support. Suppose that f lies in the domain of T, and that Tf is effectively uniformly continuous. Then Tf is computable.*

[I.e. if Tf satisfies (ii) above, then Tf also satisfies (i).]

Proof. Since the groundwork has been carefully laid, the proof is quite easy. We begin by constructing a computable sequence $\{\varphi_k\}$ of C^∞ functions which form an "approximate identity" for convolution—i.e., as $k \to \infty$, the support of φ_k approaches zero effectively, while the integral of φ_k over $\mathbb{R}^q$ remains equal to 1. In detail:

Begin with any C^∞ function $\varphi \geq 0$ with support on the unit disk in $\mathbb{R}^q$, and such that the sequence of derivatives $\{\varphi^{(\alpha)}\}$ (α = multi-index) is computable. We can assume without loss of generality that

$$\int \cdots \int_{\mathbb{R}^q} \varphi \, dx_1 \ldots dx_q = 1.$$

Let

$$\varphi_k(x) = k^q \varphi(kx).$$

Then the support of φ_k is a disk of radius $1/k$ about the origin, and the integral of φ_k over $\mathbb{R}^q$ is equal to 1. Furthermore, $\{\varphi_k^{(\alpha)}\}$ is computable, effectively in k and α. Thus $\{\varphi_k\}$ is the desired computable "approximate identity".

The following fact about convolutions is well known:

(∗) If $g: \mathbb{R}^q \to \mathbb{R}^1$ is uniformly continuous, and $\varepsilon, \delta > 0$ are chosen so that $|x - y| \leq \delta$ implies $|g(x) - g(y)| \leq \varepsilon$, then

$$1/k \leq \delta \quad \text{implies} \quad |(g * \varphi_k) - g| \leq \varepsilon.$$

Our proof consists in effectivizing this elementary fact, and at the same time using the conditions (1)–(3) to connect it to the properties of the operator T.

Since the operator T satisfies (2) and f has compact support, so does Tf.

By construction, $\{\varphi_k^{(\alpha)}\}$ is a computable multi-sequence. Since f is computable and integration is a computable process, $\{f * [\varphi_k^{(\alpha)}]\}$ is computable. Since differentiation commutes with convolution, $f * [\varphi_k^{(\alpha)}] = [f * \varphi_k]^{(\alpha)}$. Hence by (3), $\{T[f * \varphi_k]\}$ is a computable sequence.

Now we use (1)—translation invariance. This gives $T[f * \varphi_k] = [Tf] * \varphi_k$, and hence by the above $\{[Tf] * \varphi_k\}$ is a computable sequence. We will show that $[Tf] * \varphi_k$ converges effectively and uniformly to Tf, so that by Theorem 4 of Chapter 0, Tf is computable.

Here we use the assumption that Tf is effectively uniformly continuous. Thus there is a recursive function $d(N)$ such that $|x - y| \leq 1/d(N)$ implies $|Tf(x) - Tf(y)| \leq 2^{-N}$. Now the estimate (∗) above becomes effective, and we have

$$k \geq d(N) \quad \text{implies} \quad |([Tf] * \varphi_k) - Tf| \leq 2^{-N}.$$

Thus $[Tf] * \varphi_k$ approaches Tf effectively and uniformly. As we have already noted, this implies that Tf is computable. The proof of Theorem 7 is complete. □

As promised, we give two applications of Theorem 7. The first—relating to partial derivatives—is a generalization of Theorem 2 in Section 1. The second gives information about noncomputable solutions of the wave equation.

Partial derivatives. We now give conditions which ensure the computability of partial derivatives. We shall take pains to present the result in its most general form.

Let $\alpha = (\alpha_1, \ldots, \alpha_q)$ be a multi-index of order $|\alpha| = \alpha_1 + \cdots + \alpha_q$. Let D^α denote the partial differential operator

$$D^\alpha f = f^{(\alpha)} = \frac{\partial^{|\alpha|} f}{\partial x_1^{\alpha_1} \ldots \partial x_q^{\alpha_q}}.$$

Let $f: \mathbb{R}^q \to \mathbb{R}^1$ be a computable function with compact support. Suppose that all partial derivatives of f of order $< |\alpha|$ are continuous (although not necessarily computable, and not necessarily effectively uniformly continuous). Suppose that the particular derivative $D^\alpha f$ is effectively uniformly continuous. Then $D^\alpha f$ is computable.

[Of course, as a corollary, if f is C^∞, then every partial derivative $D^\alpha f$ is computable.]

To prove this, we merely observe that the operator $T = D^\alpha$ satisfies conditions (1)–(3) above. Hence the result is an immediate consequence of Theorem 7.

Note. The general hypotheses above—in which only $D^\alpha f$ itself is assumed effectively uniformly continuous—can actually occur. Example: Let $f(x)$ be as in Theorem 1 above, so that $f'(x)$ is continuous but not effectively uniformly continuous.

Let $h(x, y) = f(x)$. Then $\partial h/\partial x = f'(x)$ is not effectively uniformly continuous, but $\partial^2 h/\partial x \partial y = 0$ is.

Of course this is a technicality. However, the general result for $D^\alpha f$ above—a trivial consequence of Theorem 7—would be much harder to prove by the elementary methods used in the proof of Theorem 2.

The wave equation. In Section 3 above, we remarked that the wave equation with computable initial data can have a noncomputable solution. In fact we can start with computable initial data at time $t = 0$ and obtain a noncomputable solution at time $t = 1$. (This is further discussed in Section 5 of Chapter 3.)

4. Translation Invariant Operators

Here we consider one aspect of these noncomputable solutions—an aspect which is related to Theorem 7 above. The noncomputable solutions are always "weak solutions"—i.e. although continuous, they are not C^2 or even C^1. Indeed, if they were C^1, they would be effectively uniformly continuous. This is impossible, for we have:

Theorem 8 (Noncomputable solutions of the wave equation must be weak solutions). *Consider the wave equation*

$$\frac{\partial^2 u}{\partial x^2} + \frac{\partial^2 u}{\partial y^2} + \frac{\partial^2 u}{\partial z^2} - \frac{\partial^2 u}{\partial t^2} = 0,$$

with the initial conditions $u(x, y, z, 0) = f(x, y, z)$, $\partial u/\partial t = 0$ *at time* $t = 0$. *Suppose that f is computable and continuous with compact support. Suppose, however, that the solution $u(x, y, z, 1)$ is not computable. Then $u(x, y, z, 1)$ is not effectively uniformly continuous, and hence is not C^2 or even C^1.*

Notes. As we will show in Chapter 3, such noncomputable solutions can be continuous. Thus the break-point lies between continuity and effective uniform continuity.

Finally a trivial note: of course the time $t = 1$ could be replaced by any computable time $t = t_0$.

Proof. Suppose that $u(x, y, z, 1)$ is effectively uniformly continuous. We will show that then $u(x, y, z, 1)$ is computable.

Let T be the solution operator which maps the initial data $f(x, y, z)$ onto the solution $u(x, y, z, 1)$. If we can show that T satisfies conditions (1)–(3) above, then the desired result will follow immediately from Theorem 7.

Now there is a well known formula (Kirchhoff's equation) for T: it is displayed in Chapter 3, Section 5. Using this formula, it is a routine matter to verify that T satisfies (1)–(3). However, as an illustration of technique the following seems more interesting:

Two of the key properties of T can be seen on "physical" grounds. That T is translation invariant—condition (1) above—follows from the fact that the wave equation is translation invariant. That T preserves compact supports—condition (2) above—follows from the fact that waves travel with a finite velocity. Finally (3)—computability for C^∞ functions—does require a glance at the formula in Chapter 3, Section 5: but only long enough to verify that the formula involves integrals and partial derivatives. The exact shape of the formula is irrelevant.

Since conditions (1)–(3) above are satisfied by T, the desired result follows at once from Theorem 7. □

Part II

The Computability Theory of Banach Spaces

Chapter 2
Computability Structures on a Banach Space

Introduction

In this chapter we introduce the concept, "computability structure on a Banach space". This concept will play a fundamental role throughout the remainder of the book.

The notion of a computability structure arose in response to the following question. Which processes in analysis and physics preserve computability and which do not? We first make this question precise, and then we answer it.

How will the term "process" be understood? Among the processes of physics we expect to include are those associated with the wave equation, the heat equation, Laplace's equation and many others. Among the processes of analysis we expect to consider are Fourier series, Fourier transform and others. A moment's thought will convince the reader that the solution operator associated with each of these processes is a linear operator which maps functions into functions. Hence we will be concerned with linear operators on Banach spaces of functions. They will be our "processes".

How will the term "computable" be understood? Of course, since we will be dealing with Fourier series and transforms, solutions of the wave and heat equations etc., the appropriate notion is "computable function of a real/complex variable". The classical definition of this was spelled out in Chapter 0. In order to cover the variety of processes mentioned above, we will have to extend this definition to computability on an arbitrary Banach space. The extended definition plays a key role in the statement, proof and applications of the three principal results of this book. One of them, the *First Main Theorem*, provides corollaries which answer the question proposed above: which "processes" do and which do not preserve computability. The other two are the *Second Main Theorem* and the *Eigenvector Theorem*. These two results, together with their associated theorems in Chapter 4, determine precisely the "computability relationships" between an operator and its eigenvalues/spectrum/eigenvectors in a general setting. As is well known, linear operators on Hilbert space—more particularly the unbounded ones—are of considerable importance in physical theory.

The concept of a "computability structure on a Banach space" will be given axiomatically. This is a natural approach when one wants to encompass a variety of applications.

What concept should be axiomatized? Note that it is not sufficient to axiomatize the notion of a "computable point". The reason is obvious: topological notions are required in the solution of problems in analysis. Since a Banach space is a metric space, the topology can be given by convergent sequences. This suggests that an appropriate concept for axiomatization is "computable sequence of points". (A computable point x is a point such that the sequence $x, x, x \ldots$ is computable.) Of course, the topology can also be given by other, equivalent methods—e.g. open sets. However the sequence approach is very natural in the study of computability. Indeed one of the fundamental intuitions of recursive function theory is the notion of a "computable sequence of integers"—a sequence $a(0), a(1), \ldots$ obtained by a Turing Machine computation of the recursive function a. We will see that the concept, "computability structure on a Banach space", which is in fact an axiomatization of "computable sequence of points $x_0, x_1, x_2, \ldots$ of a Banach space", will provide a useful and flexible tool for solving problems.

Once the decision is made to axiomatize "computable sequence of points", the specific axioms follow naturally. Since a Banach space is linear, we need an axiom for linearity (Axiom 1, below). Since a Banach space is complete, we require an axiom for completeness (Axiom 2). Since a Banach space is normed, we need an axiom for the norm (Axiom 3). In fact, these are the only axioms which will be postulated. A detailed presentation of the axioms will be given in the next section.

When viewed in this light, the axioms appear to be minimal: they are just sufficient to relate the concept of computability to the basic concepts of Banach space theory. We will see later that under very general conditions which are satisfied in practical situations, the axioms are not merely minimal but also maximal. Thus it is not possible to postulate axioms which, when added to the axioms of a computability structure, give a more intuitive concept of computability.

The following definitions are both natural and useful. They are concerned with an effectivization of the notion of a separable Banach space. A computability structure $\mathscr{S}$ on a Banach space X is *effectively separable* if it contains a sequence $\{e_n\}$ whose linear span is dense in X. (Recall that the linear span of a set V consists of all finite linear combinations with real/complex coefficients of the elements of V.) Such a sequence $\{e_n\}$ is called an *effective generating set* for $\mathscr{S}$.

Although it is natural for applications to define a computability structure axiomatically, an axiomatic approach may be surprising to the expert in recursion theory. There is something "genetic" about computability. Witness, for example, the definition of computable function from $\mathbb{N}$ to $\mathbb{N}$. As is well-known, early research in this area consisted in attempting to capture the intuitive concept of computability by an appropriate definition. Definitions of this concept were presented by Turing, Post, Herbrand/Gödel, Markov, etc. The fact that the insights of Post, Turing, etc. led to the same class of functions was important in concluding that the "right" definition of "computable function from $\mathbb{N}$ to $\mathbb{N}$" had been found. Recursion theory on $\mathbb{N}$ is the study of the consequences of this definition.

By contrast, the axiomatic approach to a computability structure allows for a multitude of computability structures on a Banach space. (As mentioned earlier this is essential for applications.) Not all of these structures can be "right". Yet, surprisingly enough, the axiomatic approach appears to capture the concept of computability rather well. There are several reasons for this, which we now consider.

Introduction

Associated with each of the usual Banach spaces—e.g. $C[0, 1]$, $L^p[0, 1]$, l^p ($1 \leq p < \infty$) etc.—there is an intrinsic notion of computability which satisfies the axioms of a "computability structure". This will be discussed in greater detail later in the chapter (Sections 2, 3, 4). At this point we merely remark that the intrinsic computability structure is unique. For the space $C[0, 1]$, we will see that it is given by the "computable sequence of computable functions" defined in Chapter 0. We note further that, for each of the spaces mentioned above, intrinsic computability can be presented in many ways. The different ways originate, in part, from the different branches of analysis. This happens as follows. The specific branch of analysis focuses upon a particular, well-understood sequence of functions which is dense in the Banach space. The computability structure is obtained by taking the effective closure of this sequence. For example, if one is concerned with Fourier series, it is natural to choose the trigonometric polynomials with rational (complex rational) coefficients as the sequence and take its effective closure. If, however, one is concerned with problems in measure and integration on L^p-space ($1 \leq p < \infty$), it is more natural to consider a sequence composed of certain computable step functions—i.e. those step functions with rational jump points and values. As we will see later, each of these sequences is an example of an "effective generating set". (This concept was discussed above.) We will show that the effective closure of those effective generating sets which arise naturally from basic constructs in analysis all lead to the same computability structure—intrinsic computability. For this reason, intrinsic computability plays a fundamental role in a wide variety of applications. This fact is of considerable importance, not only for applications but also for understanding the nature of computability on a Banach space.

The genetic quality of the definition of a computability structure can also be viewed in the more general context of a separable Banach space. Recall that the axioms of a computability structure on any Banach space—whether separable or not—appear to be minimal. They are just sufficient to impact the concept of computability with the basic notions of Banach space theory—linearity, limit and norm. For an effectively separable Banach space something stronger holds. The axioms are also maximal. More specifically it is not possible to formulate axioms which, when added to the axioms of a computability structure, yield a more intuitive characterization of computability. This is a consequence of the following theorem which will appear in Section 5.

Stability Lemma. *Let $\{e_n\}$ be a sequence whose linear span is dense in the Banach space X. Let $\mathscr{S}'$ and $\mathscr{S}''$ be computability structures on X such that $\{e_n\} \in \mathscr{S}'$ and $\{e_n\} \in \mathscr{S}''$. Then $\mathscr{S}' = \mathscr{S}''$.*

Thus we see that if a computability structure $\mathscr{S}$ contains an effective generating set, then it can have no proper extension. This rigidity in the structure of $\mathscr{S}$ implies the aforementioned redundancy of any further axioms. (Note also that $\mathscr{S}$ cannot have any proper restriction containing the effective generating set.)

The fact that for an effectively separable Banach space, the axioms of a computability structure are both maximal and minimal seems to indicate that, at least for those spaces, the concept of a computability structure does capture the intuitive notion which we need—computability on a Banach space.

Although the Stability Lemma is very general, it does require that the two structures $\mathscr{S}'$ and $\mathscr{S}''$ share a common effective generating set $\{e_n\}$. This condition is satisfied in most practical cases. However, for the purpose of constructing counterexamples, we may sometimes consider computability structures which do not contain the usual effective generating sets. Such artificial computability structures do exist, and we call them "ad hoc structures".

It should be pointed out that ad hoc computability structures play an important role in our research. They are *not* useless by-products of the axiomatic approach, as we will see. One way in which they are used is as follows. There are some natural techniques in analysis—particularly for obtaining counterexamples—which are noncomputable. They lead readily to ad hoc computability structures. In some instances it is possible to translate the ad hoc computability structure into the intrinsic one. In the process of translation, the analytic intuition which led to the construction is so thoroughly disguised that, for all practical purposes, it is lost. The counterexamples would be difficult to discover without first passing through an ad hoc structure. The proof of the Eigenvector Theorem (Theorem 6, Chapter 4) provides an example of an ad hoc structure which is used in this way.

Before concluding this section, the following should be remarked. We have NOT defined a computable Banach space. We began with a preexisting Banach space and defined a computability structure on it. Furthermore, when reasoning about Banach space theory, we employ classical logic, as physicists and analysts do. In particular we make free use of nonconstructive methods of proof to characterize the set of computable sequences within the larger class of all sequences. Thus our work is firmly within the tradition of recursive function theory.

1. The Axioms for a Computability Structure

We begin with a technical note. Virtually everything we do applies mutatis mutandis to either real or complex Banach spaces. The same holds for explicit function spaces such as $C[a, b]$, $L^p[a, b]$, etc. We shall take this as being understood, and shall rarely mention it in what follows.

Let X be a Banach space. The undefined term, which the following axioms will govern, is "computable sequence $\{x_n\}$", $x_n \in X$. We denote the set of all computable sequences by $\mathscr{S}$. Thus a Banach space with a computability structure is really a pair $\langle X, \mathscr{S} \rangle$. By a conventional "abuse of notation", we shall sometimes refer merely to X (or $\mathscr{S}$), the other member of the pair being understood.

A double sequence $\{x_{nm}\}$ is called "computable" if it is mapped onto a computable sequence by one of the standard recursive pairing functions from $\mathbb{N} \times \mathbb{N}$ onto $\mathbb{N}$. Similarly for triple or k-fold sequences.

An element $x \in X$ is called "computable" if the sequence $\{x, x, x, \ldots\}$ is computable.

The following simple definition will be used not only in the axioms, but throughout the remainder of the book. It is an obvious adaptation of the notion of effective convergence (which was discussed in Chapter 0) to Banach spaces.

1. The Axioms for a Computability Structure

Definition. The double sequence $\{x_{nk}\}$ converges to the sequence $\{x_n\}$ as $k \to \infty$, effectively in k and n, if there exists a recursive function e such that

$$\|x_{nk} - x_n\| \leq \frac{1}{2^N}.$$

for all $k \geq e(n, N)$.

Now we state the axioms. The first one covers finite linear combinations.

Axiom 1 (Linear Forms). Let $\{x_n\}$ and $\{y_n\}$ be computable sequences in X, let $\{\alpha_{nk}\}$ and $\{\beta_{nk}\}$ be computable double sequences of real or complex numbers, and let $d: \mathbb{N} \to \mathbb{N}$ be a recursive function. Then the sequence

$$S_n = \sum_{k=0}^{d(n)} (\alpha_{nk} x_k + \beta_{nk} y_k)$$

is computable in X.

Axiom 2 (Limits). Let $\{x_{nk}\}$ be a computable double sequence in X such that $\{x_{nk}\}$ converges to $\{x_n\}$ as $k \to \infty$, effectively in k and n. Then $\{x_n\}$ is a computable sequence in X.

Axiom 3 (Norms). If $\{x_n\}$ is a computable sequence in X, then the norms $\{\|x_n\|\}$ form a computable sequence of real numbers.

We assume that at least one computable sequence in X exists, whence the sequence $\{0, 0, 0, \ldots\}$ is computable.

As corollaries we derive the following, which were given as axioms in the authors' papers [1983b, 1987].

Proposition. *The following are consequences of the axioms.*

(1) *(Composition property). If $\{x_n\}$ is a computable sequence in X, and $a: \mathbb{N} \to \mathbb{N}$ is a recursive function, then $\{x_{a(n)}\}$ is a computable sequence in X.*

(2) *(Insertion property). If $\{x_n\}$ and $\{y_n\}$ are computable sequences in X, then the sequence $\{x_0, y_0, x_1, y_1, x_2, y_2, \ldots\}$ is computable.*

Proofs. For (1): Let $y_n = 0$ for all n. Let $d(n) = a(n)$, and let $\{\alpha_{nk}\}$ be the computable double sequence of reals given by:

$$\alpha_{nk} = 1 \quad \text{if } k = a(n), \quad 0 \text{ otherwise.}$$

Then

$$S_n = \sum_{k=0}^{d(n)} \alpha_{nk} x_k = x_{a(n)}.$$

Hence by the Linear Forms Axiom, $\{x_{a(n)}\}$ is computable.

For (2): Let $\{x_n\}$ and $\{y_n\}$ be the given sequences. Let $d(n) = n$, and let $\{\alpha_{nk}\}$ and $\{\beta_{nk}\}$ be the computable real sequences given by:

$\alpha_{nk} = 1$ if n is even and $k = n/2$,   0 otherwise;

$\beta_{nk} = 1$ if n is odd and $k = (n-1)/2$,   0 otherwise.

Then

$$s_n = \sum_{k=0}^{d(n)} (\alpha_{nk} x_k + \beta_{nk} y_k) = \begin{cases} x_{n/2} & \text{if } n \text{ is even,} \\ y_{(n-1)/2} & \text{if } n \text{ is odd.} \end{cases}$$

Again by the Linear Forms Axiom, the interspersed sequence $\{x_0, y_0, \ldots\}$ is computable. □

We conclude this section by recalling some definitions which will play a key role in many of the results which follow.

Let $\{e_n\}$ be any sequence of vectors. We recall that the *linear span* of $\{e_n\}$ is the set of all finite (real/complex) linear combinations of the e_n. For example, if $\{e_n\}$ is the sequence of monomials $1, x, x^2, x^3, \ldots$, then the linear span of $\{e_n\}$ is the set of all polynomials.

Definition. Let X be a Banach space with a computability structure $\mathscr{S}$. We say that $\langle X, \mathscr{S} \rangle$ is *effectively separable* if there exists a computable sequence $\{e_n\}$, $e_n \in X$, such that the linear span of $\{e_n\}$ is dense in X. Such a sequence $\{e_n\}$ is called an *effective generating set* for $\langle X, \mathscr{S} \rangle$.

By an abuse of notation we shall sometimes refer to the Banach space X itself as being effectively separable.

Remark. We do not assume above that $\{e_n\}$ is "effectively dense"—merely that it is dense. Actually the effective density condition follows automatically. This will be proved in the Effective Density Lemma (Theorem 1, Section 5). The effective density condition was redundantly made part of the definition in the authors' paper [1983b].

2. The Classical Case: Computability in the Sense of Chapter 0

It is not hard to show that the computable sequences of computable functions of Chapter 0 satisfy the axioms of a computability structure. For let a and b be recursive reals, and let $C[a, b]$ be the Banach space of all continuous functions on $[a, b]$ with the uniform norm. Suppose further that $\mathscr{S}$ is the collection of all

3. Intrinsic L^p-computability

computable sequences of computable functions as defined in Chapter 0. It is trivial to verify that Axiom 1 (Linear Forms) holds for $\mathscr{S}$. Furthermore the Limit Axiom (Axiom 2) is Theorem 4, and the Norm Axiom (Axiom 3) is Theorem 7 of that chapter.

It is just as easy to prove that $C[a, b]$ is effectively separable with respect to $\mathscr{S}$. For the monomials $1, x, x^2, \ldots$ form a computable sequence whose linear span is dense in $C[a, b]$. Thus $\{x^n\}$ is an effective generating set.

Another effective generating set is the collection of all piecewise linear functions in $C[a, b]$ with rational (complex rational) coordinates for the "corners".

A third example applies to Fourier series. Here, for convenience, let $[a, b] = [0, 2\pi]$. In this section, where the norm is that of uniform convergence, we must restrict attention to continuous functions f such that $f(0) = f(2\pi)$. [In Section 3 below, where we deal with L^p, the condition $f(0) = f(2\pi)$ can, of course, be dropped.] Now the effective generating set which applies most naturally to this situation is the sequence of trigonometric functions:

$$1, \cos x, \sin x, \cos 2x, \sin 2x, \ldots.$$

It is worth noting that by Theorem 6 of Chapter 0 the effective generating set $\{x^n\}$ has a stronger property: every computable sequence $\{f_n\}$ is the *effective* limit of a computable sequence of polynomials of the form

$$p_{nk} = \sum_{j=0}^{d(n,k)} a_{nkj} x^j,$$

where $\{a_{nkj}\}$ is a computable triple sequence of rationals and d is a recursive function. We will see in Section 5 below, that this property generalizes to arbitrary effectively separable Banach spaces. For suppose $\{e_n\}$ is an effective generating set for $(X, \mathscr{S})$. Then, by the Effective Density Lemma, every computable sequence $\{x_n\} \in X$ is the effective limit of a computable sequence of polynomials

$$p_{nk} = \sum_{j=0}^{d(n,k)} a_{nkj} e_j.$$

3. Intrinsic L^p-computability

The definition of intrinsic L^p-computability is a natural generalization of computability as defined in Chapter 0.

We begin with a brief comment on notation. Let $1 \leq p < \infty$ be a computable real; let a and b be computable reals.

We now turn to $L^p[a, b]$. For $p = \infty$ we take the space $C[a, b]$ with the uniform norm. Recall that classically $C[a, b]$ is dense in $L^p[a, b]$. Thus the following definition suggests itself.

Definition. (a) A function $f \in L^p[a, b]$ is L^p-computable if there exists a sequence $\{g_k\}$ of continuous functions which is computable (in the sense of Chapter 0) and such that the L^p-norms $\|g_k - f\|_p$ converge to zero effectively as $k \to \infty$.

(b) A sequence $\{f_n\}$ of L^p-functions is L^p-computable if there exists a double sequence $\{g_{nk}\}$, which is computable (in the sense of Chapter 0) and such that $\|g_{nk} - f_n\|_p$ converges to zero as $k \to \infty$, effectively in n and k.

Computability for $L^p(I^q)$—where I^q is a computable rectangle of $\mathbb{R}^q$ (see Section 3, Chapter 0)—is defined similarly.

It is not hard to verify that the computable sequences of $L^p[a, b]$ defined above satisfy the axioms of a computability structure. Furthermore, this structure is effectively separable. Three effective generating sets are inherited from the computability structure on $C[a, b]$. They are: the monomials $1, x, x^2, \ldots$; the collection of all piecewise linear functions in $C[a, b]$ with rational coordinates for the "corners"; and the trigonometric polynomials. (See the preceding section.)

There is, however, a fourth effective generating set for $L^p[a, b]$, with $p \neq \infty$. This is the collection of all step functions with rational values and jump points. We observe that step functions could not have been used in conjunction with $C[a, b]$, since they are not continuous. In fact, it has been part of the accepted "folk wisdom" of this subject that a discontinuous function cannot be computable. This holds, however, only so long as we view the computation of a function as being carried out pointwise. Now, for L^p-functions, pointwise evaluation is not well-defined, since an L^p-function is determined only almost everywhere. By contrast, for functions in $C[a, b]$, where we use the uniform norm, pointwise evaluation makes perfectly good sense. Indeed uniform convergence implies pointwise convergence.

We now turn to noncompact domains. For $p \neq \infty$, $L^p(\mathbb{R})$ is our prototype. For $p = \infty$ our space is $C_0(\mathbb{R})$, the collection of all continuous functions which approach zero as $|x| \to \infty$.

The definition of L^p-computability on $\mathbb{R}$ is an extension of the definition of L^p-computability on $[a, b]$. We require that the functions g_k and g_{nk} have compact supports which vary effectively in n and k. Without loss of generality, we can assume that the kth function is supported on a disk of radius k. This gives

Definition. A sequence $\{f_n\} \in L^p(\mathbb{R})$ is computable if there exists a sequence $\{g_{nk}\}$, computable in the sense of Chapter 0, such that
 (a) the support of g_{nk} is included in $[-k, k]$,
 (b) $\|f_n - g_{nk}\|_p \to 0$ as $k \to \infty$, effectively in n and k.

Computability for $L^p(\mathbb{R}^q)$ is defined similarly.

Note. Suppose $p = \infty$. If $f \in C[a, b]$, then the first definition gives the computable functions and sequences on $C[a, b]$ as defined in Chapter 0. If $f \in C_0(R)$, then the second definition shows that f is computable if and only if f is computable as in Section 3 of Chapter 0, and $f(x) \to 0$ effectively as $|x| \to \infty$.

Once again it is easy to see that intrinsic computability on $L^p(\mathbb{R})$, $1 \leq p \leq \infty$, satisfies the axioms of a computability structure. Furthermore, the computability structure is effectively separable. Note however, that the monomials do not form an effective generating set, as $x^n \notin L^p(\mathbb{R})$. Similarly for trigonometric polynomials. However if we take those continuous functions with compact support which are piecewise linear and have rational coordinates at their "corners", then we obtain an effective generating set. For $p \neq \infty$, the step functions with rational jump points and values also are an effective generating set.

4. Intrinsic l^p-computability

As in functional analysis, let l^p (for $1 \leqslant p < \infty$) denote the space of all real (or complex) sequences $\{c_k\}$ such that the norm $(\Sigma |c_k|^p)^{1/p}$ is finite. Of course, we assume that p is a computable real. For $p = \infty$ (which corresponds to the sup norm), our space is l_0^∞, the space of all sequences $\{c_k\}$ which converge to zero as $k \to \infty$.

Definition. (a) Let $\{c_k\} \in l^p$. Then $\{c_k\}$ is l^p-computable if $\{c_k\}$ is a computable sequence of real or complex numbers and $\Sigma |c_k|^p$ converges effectively.

Let $\{c_k\} \in l_0^\infty$. Then $\{c_k\}$ is l_0^∞-computable if $\{c_k\}$ is a computable sequence of real or complex numbers which converges effectively to zero as $k \to \infty$.

(b) Let $\{x_n\}$ be a sequence of elements in l^p (i.e. $x_n = \{c_{nk}\}$, where each c_{nk} is a number). Then $\{x_n\}$ is l^p-computable if (i) $\{c_{nk}\}$ is a computable double sequence of numbers, (ii) for each n, $\Sigma_k |c_{nk}|^p$ converges effectively in both k and n. Similarly if $\{x_n\}$ is a sequence of l_0^∞, then $\{x_n\}$ is l_0^∞-computable if $\{c_{nk}\}$ is a computable sequence, such that $\{c_{nk}\}$ converges to zero as $k \to \infty$, effectively in k and n.

It is simple to verify that the axioms of a computability structure hold for these theories. Furthermore all of these spaces are effectively separable. As an effective generating set, we can take the sequence of vectors $e_n = \{0, 0, 0, \ldots, 0, 1, 0, 0, \ldots\}$ with a 1 in the nth place.

5. The Effective Density Lemma and the Stability Lemma

In this section we prove two basic lemmas. The first of these, the Effective Density Lemma, states roughly that an effective generating set is not merely dense but "effectively dense". This fact is quite useful. It allows us to conclude without further proof, that certain (classical) approximation theorems have effective versions. For example, by the (classical) Weierstrass approximation theorem, the polynomials with rational coefficients are dense in $C[0, 1]$. Since these polynomials form an effective generating set, we know by the Effective Density Lemma that the Effective Weierstrass Theorem (Theorem 6 of Chapter 0) holds. The reader may recall the long and somewhat messy proof given in that chapter.

The second lemma, the Stability Lemma, was discussed briefly in the introduction to this chapter. As a consequence of this lemma, we have a kind of "uniqueness" for computability structures on a Banach space. The lemma actually shows that the specification of an effective generating set gives a computability structure (= collection of computable sequences) which has no proper extension or restriction. Thus it is not possible to invent new axioms which, together with the axioms of a computability structure, give rise to a more intuitive notion of computability.

The Stability Lemma is an immediate corollary of the Effective Density Lemma.

Theorem 1 (Effective Density Lemma). *Suppose $\{e_n\}$ is an effective generating set for X. Then the sequence $\{x_n\}$, $x_n \in X$, is computable if and only if there is a computable double sequence p_{nk} such that*

(i) $$p_{nk} = \sum_{j=0}^{d(n,k)} \alpha_{nkj} e_j,$$

where $\{\alpha_{nkj}\}$ is a computable triple sequence of rationals/complex rationals, and $d(n, k)$ is a recursive function—

(ii) $\qquad p_{nk} \to x_n$ *as* $k \to \infty$, *effectively in k and n.*

Proof. The "if" part follows immediately from the Linear Forms Axiom and the Limit Axiom. To prove the "only if" part we argue in the following way. Let $\{p_i\}$ be an effective listing of all finite linear combinations of the e_n with rational (complex rational) coefficients. By the Linear Forms Axiom, $\{p_i\}$ is computable in X. Thus $p_i - x_n$ is a computable double sequence, and by the Norm Axiom, $\{\|p_i - x_n\|\}$ is computable in $\mathbb{R}$. But the linear span of $\{e_n\}$ is dense in X. So there exists an i such that $\|p_i - x_n\| < 2^{-k}$. In order to compute p_{nk}, fix x_n and generate the p_i until one satisfying $\|p_i - x_n\| < 2^{-k}$ is found. Then let $p_{nk} = p_i$. Since the index $i = i(n, k)$ is a recursive function of n and k, we have, by the Composition Property (see Section 1 of this chapter) that $\{p_{nk}\}$ is computable in X. $\square$

Corollary 1a (Effective Weierstrass Theorem). *Let I^q be a q-dimensional rectangle with computable coordinates for the "corners". Let X be the Banach space $C(I^q)$ with the uniform norm. Then $f \in C(I^q)$ is computable if and only if there is a computable sequence of polynomials $p_k(x_1, \ldots, x_q)$ with rational coeffficients which converges uniformly to f, effectively in k.*

Similarly $\{f_n\}$ is a computable sequence of $C(I^q)$ if and only if there is a computable double sequence $p_{nk}(x_1, \ldots, x_q)$ with rational coefficients which converges uniformly to f_n as $k \to \infty$, effectively in k and n.

Proof. An immediate corollary of the Effective Density Lemma. The sequence of monomials $\{x_1^{n_1}, x_2^{n_2}, \ldots, x_q^{n_q}\}$ is an effective generating set for $C(I^q)$. $\square$

5. The Effective Density Lemma and the Stability Lemma

As deeper applications in the same vein, we give effective versions of two famous approximation theorems: the Stieltjes, Hamburger, Carleman Theorem and the Wiener Tauberian Theorem.

For the first of these: fix a computable real number $\alpha > 0$. Let $G^\alpha(\mathbb{R})$ denote the Banach space of all continuous function f on $(-\infty, \infty)$ such that $\lim_{x \to \pm\infty} f(x)e^{-|x|^\alpha} = 0$, endowed with the norm

$$\|f\| = \sup_x |f(x)|e^{-|x|^\alpha}.$$

We put a computability structure on $G^\alpha(\mathbb{R})$ by defining a sequence of continuous functions $\{f_n\}$ to be *computable* if:

(i) the sequence $\{f_n\}$ is computable in the sense of Chapter 0 and

(ii) $\lim_{x \to \pm\infty} f_n(x)e^{-|x|^\alpha} = 0$, effectively as $x \to \pm\infty$ and effectively in n.

Corollary 1b (Effective Stieltjes, Hamburger, Carleman Theorem). *Let α be a computable real, $\alpha \geqslant 1$. Let f be a continuous function on $(-\infty, \infty)$ which is computable in $G^\alpha(\mathbb{R})$. Then there exists a computable sequence of polynomials which converges effectively to f in the norm $\|h\| = \sup_x |h(x)|e^{-|x|^\alpha}$.*

Proof. If we ignore the question of effective convergence, this is a famous classical theorem (cf. Shohat, Tamarkin [1943]). But the effectiveness of the convergence follows immediately from the Effective Density Lemma.

Similarly, and with no extra effort, we obtain:

Corollary 1c (Effective Wiener Tauberian Theorem). *Let g be a computable function in $L^1(\mathbb{R})$ whose Fourier transform $\hat{g}(t) \neq 0$ for all real t. Let f be an arbitrary computable function in $L^1(\mathbb{R})$. Then there exists a computable sequence of rational linear combinations of translates of g which converges effectively to f in L^1 norm.*

Proof. An immediate consequence of the classical (noneffective) Wiener Tauberian Theorem, together with the Effective Density Lemma.

Theorem 2 (Stability Lemma). *Let $\{e_n\}$ be a sequence whose linear span is dense in the Banach space X. Let $\mathscr{S}'$ and $\mathscr{S}''$ be computability structures on X such that $\{e_n\} \in \mathscr{S}'$ and $\{e_n\} \in \mathscr{S}''$. Then $\mathscr{S}' = \mathscr{S}''$.*

Proof. This is an immediate consequence of the Effective Density Lemma. For that lemma applies to *any* computability structure $\mathscr{S}$. It gives—in terms of the sequence $\{e_n\}$—a precise characterization of the collection of computable sequences $\{x_n\} \in \mathscr{S}$. □

As a corollary we obtain the following proposition.

Corollary 2a. *Let $\{e_n\}$ be a sequence whose linear span is dense in the Banach space X. Let $\mathscr{S}, \mathscr{S}'$ be computability structures on X such that $\{e_n\} \in \mathscr{S}'$ and $\mathscr{S}' \subseteq \mathscr{S}$. Then $\mathscr{S}' = \mathscr{S}$.*

6. Two Counterexamples: Separability Versus Effective Separability and Computability on $L^\infty[0, 1]$

Throughout this book, the notion of effective separability will play an important role. Since most of the Banach spaces we deal with are obviously separable, one might wonder whether the hypothesis of effective separability is redundant. Our first example shows it is not redundant: we exhibit a Banach space with a computability structure which is separable but not effectively separable.

The second example involves a Banach space, $L^\infty[0, 1]$, which has no natural computability structure at all.

Recall that a Banach space X is effectively separable if it contains an effective generating set $\{e_n\}$—i.e. a computable sequence $\{e_n\}$ whose linear span is dense in X.

Example 1. There exists a Hilbert space H with a computability structure $\mathscr{S}$ such that:

(1) The space H is separable.
(2) The computable elements of H are dense in H.
(3) The pair $\langle H, \mathscr{S} \rangle$ is not effectively separable.

We mention that condition (2) above eliminates trivial and inconsequential examples such as the following: Take a closed subspace H_0 of H, and define "computability" only for sequences $\{x_n\}$ in H_0. Conditions (2) and (3) together mean that there is something "seriously wrong" with the computability structure $\mathscr{S}$.

Proof. Now we develop the example. For H we take the Hilbert space $L^2[0, 2\pi]$. It is the computability structure $\mathscr{S}$ which is nonstandard. In brief, we shall use a definition of "computability" which—compared to the standard definition—is too stringent.

We define a sequence of functions $\{f_n\}$ in $L^2[0, 2\pi]$ to be "ultra-computable" if and only if $\{f_n\}$ is a sequence of trigonometric polynomials of *uniformly bounded degrees* (and, of course, with computable coefficients). More precisely, we say that $\{f_n\}$ is "ultra-computable" if, for some integer N (which may depend on the sequence $\{f_n\}$):

$$f_n(x) = \sum_{m=-N}^{N} c_{nm} e^{imx},$$

where $\{c_{nm}\}$ is a computable double sequence of complex numbers.

Ultra-computability satisfies all of the axioms for a computability structure on a Banach space. The Linear Forms Axiom is obvious, since any linear combination of trigonometric polynomials of degree $\leq N$ is again a trigonometric polynomial of degree $\leq N$. The Norm Axiom is also obvious, since "ultra-computability" implies "ordinary L^2-computability", for which we know that the Norm Axiom holds.

The Limit Axiom is a little harder. We must show that if $\{f_{nk}\}$ is ultra-computable and converges effectively in L^2-norm to $\{f_n\}$ as $k \to \infty$, then $\{f_n\}$ is ultra-computable. Now by definition, if $\{f_{nk}\}$ is ultra-computable, then the functions f_{nk} are all

6. Two Counterexamples: Separability Versus Effective Separability

trigonometric polynomials of degree $\leq N$. This is because a double sequence $\{f_{nk}\}$ is computable if and only if the single sequence $\{h_j\}$ given by $h_{j(n,k)} = f_{nk}$ is computable —where j is a standard pairing function from $\mathbb{N} \times \mathbb{N}$ onto $\mathbb{N}$. It is now easy to see that the limit sequence $\{f_n\}$ is also a sequence of trigonometric polynomials of degree $\leq N$. For a function $g \in L^2[0, 2\pi]$ is a trigonometric polynomial of degree $\leq N$ if and only if the inner products

$$(g, e^{ipx}) = 0$$

for all p with $|p| > N$. Now by Schwarz' inequality (i.e. $|(g, h)| \leq \|g\| \|h\|$) we deduce: If the condition $(g, e^{ipx}) = 0$ holds for $g = f_{nk}$, then since $f_{nk} \to f_n$ in L^2-norm, the condition $(g, e^{ipx}) = 0$ also holds for the limits $g = f_n$. Hence the f_n are trigonometric polynomials of degree $\leq N$, as desired.

It is trivial to verify that the sequence of coefficients $\{c_{nm}\}$ of $\{f_n\}$ is computable.

Of course the space $L^2[0, 2\pi]$ is separable. It is easy to see that, in terms of ultra-computability, this space is not effectively separable. That is, there is no ultra-computable effective generating set $\{e_n\}$. Indeed any ultra-computable sequence $\{e_n\}$ must, by definition, consist only of trigonometric polynomials of degree $\leq N$. Then the closed linear span of $\{e_n\}$ will still consist only of trigonometric polynomials of degree $\leq N$, and cannot give all of $L^2[0, 2\pi]$.

Finally, the ultra-computable elements form a dense subset of $L^2[0, 2\pi]$, since every trigonometric polynomial with complex-rational coefficients is ultra-computable. □

Remarks. The preceding paragraph harks back to Chapter 0, Section 2, where we already saw a clear distinction between "pointwise computability" (i.e. computability of the individual elements of a sequence) and "sequential computability" (which is a global statement involving the sequence as a whole). The notion of ultra-computability used above gave a wide class of single computable elements: namely all trigonometric polynomials with computable coefficients. But the definition of sequential computability, requiring a *fixed* degree N (or less) for all of the elements of the sequence, was ultra-severe. We could not have an effective generating set which was ultra-computable, because the restriction to a fixed bound on the degrees made it impossible to span all of $L^2[0, 2\pi]$.

We turn now to $L^\infty[0, 1]$, the space of all bounded measurable functions on $[0, 1]$. This space is not separable at all. We shall show that it has no "natural" computability structure. To do this, we define condition (C) as follows:

(C): If $\{a_n\}$ and $\{b_n\}$ are computable sequences of reals with $a_n < b_n$, then the sequence of characteristic functions $\chi_{[a_n, b_n]}$ is computable.

As noted in Section 3, condition (C) holds in $L^p[0, 1]$ where $1 \leq p < \infty$ and p is computable. However it must fail for $L^\infty[0, 1]$, as the following example shows.

Example 2. There is no computability structure on $L^\infty[0, 1]$ which satisfies condition (C).

Proof. Let $d: \mathbb{N} \to \mathbb{N}$ be a $1-1$ recursive function enumerating a recursively enumerable nonrecursive set D. Let $a_n = 0$ for all n. Let

$$b_n = \begin{cases} \dfrac{1}{2} & \text{if } n \notin D, \\ \dfrac{1}{2} + \dfrac{1}{2^m} & \text{if } n = d(m), m \geq 1, \\ 1 & \text{if } n = d(0). \end{cases}$$

This description of b_n, as given, is not effective. However, $\{b_n\}$ is a computable sequence of reals. For we can approximate b_n—effectively—to within an error of 2^{-N} by computing $d(m)$ for $0 \leq m \leq N$. If one of these values $d(m) = n$, then we know b_n exactly: $b_n = (1/2) + (1/2^m)$ (or $b_n = 1$ if $m = 0$). If $d(m) \neq n$ for $0 \leq m \leq N$, then the value $1/2$ approximates b_n to within an error $< 2^{-N}$.

Suppose that there were a computability structure on $L^\infty[0, 1]$ which satisfied (C). Then $\{\chi_{[a_n, b_n]}\}$ would be computable. Similarly we would have that $\chi_{[1/2, 1]}$ is computable. Thus, by the Linear Forms Axiom, $\{\chi_{[a_n, b_n]} + \chi_{[1/2, 1]}\}$ would be computable.

The L^∞ norms of this sequence are:

$$\begin{array}{ll} 2 & \text{if } n \in D, \\ 1 & \text{if } n \notin D. \end{array}$$

By the Norm Axiom this sequence should be computable. Clearly it is not. $\square$

7. Ad Hoc Computability Structures

In Sections 2, 3 and 4 we defined the "natural" computability structures on $C[a, b]$, L^p and l^p. These structures are called *intrinsic*. Moreover, in Section 5 we showed that an arbitrary computability structure (i.e. any structure satisfying the axioms) must coincide with the intrinsic structure, provided only that the two structures possess a computable dense sequence in common.

The above-mentioned results lead to the following questions.

1) Do non-intrinsic computability structures exist?
2) Do non-intrinsic computability structures—if they exist—serve any useful purpose?
3) What is the relation between the intrinsic and non-intrinsic computability structures?

This section is conerned with the answer to 1). In later chapters we shall answer 2) and 3). The answers are respectively: 1) Yes, non-intrinsic structures do exist;

7. Ad Hoc Computability Structures

2) they are useful for the creation of counterexamples (Chapter 4, Sections 5 and 6); 3) the non-intrinsic structures are isomorphic to the intrinsic one for Hilbert space, but not for Banach spaces in general (Chapter 4, Sections 6 and 7).

We now turn to the first question. We give two simple examples of ad hoc computability structures. We note in passing that both of these structures are effectively separable.

Example 3. (An ad hoc computability structure for $C[0, 1]$).

Let $C[0, 1]$ be the class of continuous *complex-valued* functions on $[0, 1]$. Let $\mathscr{S}_0$ be the intrinsic computability structure defined in Section 2. Then there exists an ad hoc computability structure $\mathscr{S}_1$ on $C[0, 1]$ such that $\mathscr{S}_0$ and $\mathscr{S}_1$ have only one sequence in common—the sequence $\{0, 0, 0, \ldots\}$.

Proof. Let c be a noncomputable complex number such that $|c| = 1$. (The existence of such a number follows from cardinality arguments, since there are uncountably many numbers on the unit circle.) The computability structure $\mathscr{S}_1$ is defined as follows. The sequence $\{f_n^c\}$ is computable in $\mathscr{S}_1$ if there exists a computable sequence $\{f_n\}$ of $\mathscr{S}_0$ such that

$$f_n^c = cf_n.$$

It is trivial to show that the sequences $\{f_n^c\}$ satisfy the axioms of a computability structure. We now show that $\mathscr{S}_0 \cap \mathscr{S}_1 = \{0, 0, 0, \ldots,\}$.

Suppose there exists a function $g^c \neq 0$ which is an element of a sequence in $\mathscr{S}_0$. Then $g^c(\xi) \neq 0$ for some computable real ξ. Now since $g^c(x) = cg(x)$ for some $g \in \mathscr{S}_0$, we have $g^c(\xi) = cg(\xi)$. Since both g and g^c are members of $\mathscr{S}_0$, $g(\xi)$ and $g^c(\xi)$ are both computable complex numbers. Hence c is computable, a contradiction. □

The previous example involved complex-valued functions. It is possible to obtain ad hoc structures for real Banach spaces. The following example shows how translation of the domain can lead to such structures.

Example 4. (An ad hoc computability structure for $C_0(R)$.)

Let $C_0(R)$ be the space of all continuous real-valued functions on R which approach 0 as $|x| \to \infty$. Let $\mathscr{S}_0$ be the intrinsic computability structure for $C_0(R)$ which is described in Section 3. Then there exists a computability structure $\mathscr{S}_1$ for $C_0(R)$ so that $\mathscr{S}_0 \neq \mathscr{S}_1$.

Proof. Let α be a noncomputable real. Define $\mathscr{S}_1$ as follows. The sequence $\{f_n^\alpha\}$ is computable in $\mathscr{S}_1$ if there exists a computable sequence $\{f_n\}$ of $\mathscr{S}_0$ such that

$$f_n^\alpha(x) = f_n(x + \alpha).$$

It is obvious that the sequences $\{f_n^\alpha\}$ satisfy the axioms of a computability structure. Furthermore $\mathscr{S}_1$ contains the sequence $\{g_n\}$, where

$$g_n(x) = e^{-(x+\alpha)^2} \quad \text{for all } n,$$

which is not in $\mathscr{S}_0$. □

In chapters 4 and 5 we shall study the computability of eigenvalues and eigenvectors. We will find that, in general, the eigenvalues are computable, but the eigenvectors are not. The proof of the result for eigenvectors leans heavily on an ad hoc computability structure.

Chapter 3
The First Main Theorem and Its Applications

Introduction

In this chapter we present the First Main Theorem. This theorem gives a criterion for determining which processes in analysis and physics preserve computability and which do not. A major portion of this chapter is devoted to applications.

Here by the term "process" we mean a linear operator on a Banach space. The Banach space is endowed with a computability structure, as defined axiomatically in Chapter 2. Roughly speaking, the First Main Theorem asserts:

> bounded operators preserve computability,
> and unbounded operators do not.

Although this already conveys the basic idea, it is useful to state the theorem with a bit more precision. The theorem involves a closed operator T from a Banach space X into a Banach space Y. We assume that T acts effectively on an effective generating set $\{e_n\}$ for X. Then the conclusion is: T maps every computable element of its domain onto a computable element of Y if and only if T is bounded.

We observe that there are three assumptions made in the above theorem: that T be closed, bounded or unbounded as the case may be, and that T acts effectively on an effective generating set. We now examine each of these assumptions in turn.

Consider first the assumption that T be bounded/unbounded. This assumption is viewed classically; it has no recursion-theoretic content. In this respect, the approach given here represents a generalization and unification of that followed in Chapters 0 and 1. In Chapters 0 and 1, we gave explicit recursion-theoretic codings —a different one for each theorem. The First Main Theorem also involves a coding, but this coding is embedded once and for all in the proof. In the applications of the First Main Theorem, no such coding is necessary. Thus, in these applications, we are free to regard the boundedness or unboundedness of the operator as a classical fact, and the effective content of this fact becomes irrelevant.

Consider next the assumption that T be closed. The notion of a closed operator is standard in classical analysis. It is spelled out, with examples, in Section 1. However, if the reader is willing to assume the standard fact—that all of the basic operators of analysis and physics are closed—he or she could simply skip Section 1.

We turn now to the final assumption: that T acts effectively on an effective generating set. This is the only assumption of effectiveness made in the First Main Theorem. We recall from Chapter 2 that an effective generating set is a computable sequence $\{e_n\}$ whose linear span is dense in X. Thus the notion of an effective generating set effectivizes the notion of separability. However, it does more than that. Associated with many of the basic processes of analysis and physics, there is a special class of functions which is intrinsically identified with the process being studied. For example, if the process is Fourier series on $[0, 2\pi]$, then the special class of functions is the sequence $\{e^{inx}\}$, $n = 0, \pm 1, \pm 2, \ldots$. If the process is differentiation, then it is natural to take as our special class of functions the monomials 1, x, x^2, x^3, ... Of course these are effective generating sets. Now, in applying the First Main Theorem, we have complete freedom in our choice of the effective generating set. So naturally we choose the set $\{e_n\}$ to be the special set of functions which are tailor-made for the problem at hand.

In summary, each of the three assumptions made in the First Main Theorem is generally easy to verify in applications. As we have noted, two of these assumptions —boundedness/unboundedness and closure—are considered classically, and here the results are usually well known. The effectiveness assumption—that T act effectively on an effective generating set—is also easy to verify. In fact, at times it is so easy that it is almost comical. For example, consider the case of Fourier series. Here the effective generating set is the sequence $\{e^{inx}\}$, $n = 0, \pm 1, \pm 2, \ldots$. What must we verify? Simply that the Fourier series of the functions e^{inx} can be computed effectively!

We turn now to a summary of the sections in this chapter.

Section 1 contains a discussion of closed operators. This section has no recursion-theoretic content. As noted above, a reader could skip this section without loss of continuity. However, there is one point which may be worth remarking. An unbounded closed operator is not everywhere defined. In all cases, its domain is a proper subset of the underlying Banach space.

Section 2 contains the statement and proof of the First Main Theorem.

In Section 3 we begin our discussion of applications. These are of two types. First we treat the computability theory of continuous functions, i.e. Chapter-0-computability. We begin by rederiving some results about integration and differentiation which were proved by direct methods in Chapters 0 and 1. The new proofs, based on the First Main Theorem, are very easy (cf. Theorem 1). A more instructive example is Theorem 2, about the convergence of Fourier series. Its proof requires the introduction of a Banach space slightly more complicated than most of those considered up to now. Nevertheless, this proof is also quite easy—much easier than a direct proof in the style of Chapter 0. (Such a direct proof appears in Pour-El, Richards [1983a].)

The second type of application in Section 3 concerns the relation between L^p and L^r-computability for $p \neq r$. For example, there exist functions in $L^p \cap L^r$ which are L^p-computable, but not L^r-computable. Perhaps more striking is the existence of a Chapter-0-computable function f on $\mathbb{R}$, such that $f \in L^r(\mathbb{R})$, but f is not computable in $L^r(\mathbb{R})$. Similar results are given for l^p and l^r. All of these results are spelled out in Theorem 3.

Section 4 deals with applications which are a bit more sophisticated than those given in Section 3. It begins with a complete treatment of the computability theory of Fourier series and transforms. As special cases we derive effective versions of the Plancherel Theorem and the Riemann-Lebesgue Lemma. However it is not always true that the Fourier operations effectivize. There are situations where the Fourier series/transform of a computable function is not computable. Theorem 4 lays out the complete details for Fourier series and transforms. One could find many similar applications, involving operators other than the Fourier transform. We have chosen the case of Fourier series and transforms as a prototype. The same method would apply to any linear operator for which we know the cases of boundedness and unboundedness.

Section 4 continues with a discussion of "well understood functions". Let us recall that the L^p-computable functions on $[a, b]$ are simply the effective closure in L^p-norm of the Chapter-0-computable functions. It turns out that certain step functions are L^p-computable. We emphasize that the computability of these step functions is not obtained by an ad-hoc addition of them to the list of Chapter-0-computable functions; it is an automatic consequence of the classical definition of computability once we relax the norm. Suppose we investigate this phenomenon from a broader viewpoint—that of "well understood functions". By this we mean continuous functions, piecewise linear functions, and step functions. We ask, for each of these classes of functions, precisely which ones are L^p-computable. For step functions it turns out to be just the ones we would expect—those with computable values and jump points (Theorem 5). A similar result holds for piecewise linear functions. However, for continuous functions, the result is different. There are continuous functions which are L^p-computable, but not Chapter-0-computable. The same question is asked for sequences of functions. Here the results turn out to be different.

Section 5 deals with applications to mathematical physics. There are many possible applications, of which we have selected three. We consider in turn the wave equation, the heat equation, and the potential equation. The wave equation is dealt with in two different contexts: Chapter-0-computability and computability in the energy norm. For the classical Chapter 0 notion of computability, we obtain an example of the wave equation with computable initial data such that its unique solution at time $t = 1$ is not computable (Theorem 6). This was done by direct methods in Pour-El/Richards [1981]. By contrast, when we use a norm which is more closely associated with the wave operator—the so called energy norm—then we obtain opposite results. In terms of the energy norm, solutions of the wave equation are computable (Theorem 7). Thus as already illustrated in several previous instances, the question of whether or not an operator preserves computability depends on the norm—and hence the Banach space—being considered.

By contrast, for the heat equation and potential equation, Chapter-0-computability *is* preserved. This is proved in Theorems 8 and 9.

Remark on the Dirichlet norm and Sobolev spaces. The energy norm treated in Section 5 is an example of a mixed norm—L^∞ in the time variable, and L^2 is the space variables. Moreover, the L^2-norm in the space variables is on the first derivatives, and not on the function itself. Such a norm, which is L^2 on the first derivatives, is called a Dirichlet norm. An important generalization of the Dirichlet norm is the Sobolev norm, associated with the Sobolev space $W^{k,p}$. This is a generalization in two respects. First, it involves all mixed partial derivatives of order $\leq k$. Second, the L^p-norm replaces the L^2-norm. The intrinsic computability structure for Sobolev spaces is defined in a manner exactly analogous to that which we use for the Dirichlet norm in Section 5.

1. Bounded Operators, Closed Unbounded Operators

Since this book is written for a mixed audience—i.e. logicians and analysts—we begin by reviewing some basic definitions and facts.

Let X and Y be Banach spaces and $T: X \to Y$ a linear operator from X into Y. The notions "T is bounded" and "T is closed" will play a key role in everything which follows. Consequently we review these notions in some detail.

Definition. A linear operator $T: X \to Y$ is called *bounded* if there is a constant $M \geq 0$ such that

$$\|Tx\| \leq M \cdot \|x\| \qquad \text{for all } x \in X.$$

(The smallest such M is, of course, called the *norm* of the operator T.)

In many cases, the following equivalent formulation is useful: T is bounded if and only if

$$x_n \to x \text{ in } X \quad \text{implies} \quad Tx_n \to Tx \text{ in } Y.$$

[Here and throughout, "$x_n \to x$" means that the norm $\|x_n - x\| \to 0$.]

We observe that the conclusion $Tx_n \to Tx$ can be broken into two halves:

i) Tx_n converges (to some limit y),
ii) the limit $y = Tx$.

Now we come to *closed* operators. Here the domain of T is usually not X but a subspace $\mathscr{D}(T)$ which is dense in X. (More on this below.)

Definition. A linear operator $T: \mathscr{D}(T) \to Y$ is called *closed* if, for $x_n \in \mathscr{D}(T)$:

$$x_n \to x \text{ in } X \text{ and } Tx_n \to y \text{ in } Y \quad \text{implies} \quad x \in \mathscr{D}(T) \text{ and } Tx = y.$$

Thus the contrast between "closed" and "bounded" operators lies in the placement of condition (i) above (Tx_n converges). With bounded operators, (i) is a conclusion of the assumption that $x_n \to x$. By contrast, with closed operators, (i) is part of the hypotheses.

1. Bounded Operators, Closed Unbounded Operators

As we shall see, all of the usual operators of analysis and physics are closed, although the most interesting ones are unbounded.

The above definition of a closed operator is easily seen to be equivalent to the following.

A linear operator $T: \mathscr{D}(T) \to Y$ is *closed* if and only if the domain $\mathscr{D}(T)$ is dense in X, and the graph of T [= the set of ordered pairs (x, Tx), $x \in \mathscr{D}(T)$] is a closed subset of the cartesian product $X \times Y$.

Now we return to a discussion of the domain $\mathscr{D}(T)$. The most striking fact about unbounded operators is that their domains are not the whole Banach space X but a proper dense subspace $\mathscr{D}(T)$ of X. Why is such a weird notion important in analysis and physics? The reason, simply, is that many important operators such as d/dx are only defined for a limited class of functions (differentiable functions). One might ask, why not take the differentiable functions as our Banach space? The difficulty here is that the differentiable functions are not complete in the most useful Banach space norms (e.g. the L^2-norm or the L^∞-norm). When one takes the completion (e.g. in L^2 or L^∞), one obtains the spaces $X = L^2[\ ]$ or $X = C[\ ]$ respectively. Then the domain $\mathscr{D}(T)$ of differentiable functions is a proper dense subset of the completed space $L^2[\ ]$ or $C[\ ]$.

In summary: the consideration of domains $\mathscr{D}(T) \neq X$ is forced on us by the desire to treat unbounded operators such as d/dx which are important in analysis and physics.

Our assumptions require that, even if $\mathscr{D}(T)$ is a proper subspace of X, at least $\mathscr{D}(T)$ is dense in X. This is easily achieved in all practical cases. Ignoring details, which vary slightly from case to case, the general idea is this. In the standard Banach spaces, such as $X = L^2[\ ]$ or $X = C[\ ]$, there are subspaces consisting of "very nice" functions which are dense in X. Depending on the situation, these "very nice" functions might be C^∞ functions, polynomials, trigonometric polynomials, etc. The standard unbounded operators of analysis and physics operate on these "very nice" functions. So in practice, there is no difficulty in making the domain $\mathscr{D}(T)$ dense in X.

Of course, if $\mathscr{D}(T) \neq X$ but $\mathscr{D}(T)$ is dense in X, then $\mathscr{D}(T)$ cannot be closed. Consider now an unbounded closed operator whose domain $\mathscr{D}(T) \neq X$. Then the graph of T is closed, but the domain of T is not. This is no contradiction. For, since T is unbounded, we can have $x_n \to x$ in X (with $x_n \in \mathscr{D}(T)$) but $\{Tx_n\}$ not convergent in Y. Then the points (x_n, Tx_n) on the graph of T approach no limit, and so the hypothesis of a closed graph implies nothing about this sequence. By contrast, if T is bounded (so that $x_n \to x$ *does* imply $Tx_n \to Tx$), then the hypothesis of a closed graph implies a closed domain. As noted above, since $\mathscr{D}(T)$ is dense in X, this means that $\mathscr{D}(T) = X$.

When, in general, does $\mathscr{D}(T) = X$? We have seen that this holds when T is bounded. In fact, for closed operators, the converse is true: If $\mathscr{D}(T) = X$ and T has a closed graph, then T is bounded. This is the well known Closed Graph Theorem. Thus, in summary: For a closed operator T, $\mathscr{D}(T) = X$ if and only if T is bounded.

Remark on notation. For an operator $T: X \to Y$ whose domain $\mathscr{D}(T) \neq X$ we should, if we were properly pedantic, write $T: \mathscr{D}(T) \to Y$. However, it is conventional to write $T: X \to Y$ (with the domain $\mathscr{D}(T)$ being understood), and we shall follow this convention.

It is perhaps time to give an example illustrating these concepts. The following gives two closed operators T_1, T_2, the second a proper extension of the first, and yet both of them having domains which are dense in $X = L^2[0, 1]$. Of course, as we have seen, such behavior can occur only when the operators are unbounded.

Example. Let $X = Y = L^2[0, 1]$, and let T_1, T_2 be the operators given formally by d/dx, with the domains $\mathscr{D}(T_i)$, $i = 1, 2$, defined as follows:

We begin with the class of functions $f(x)$ on $[0, 1]$ such that f is absolutely continuous (whence $f'(x)$ exists almost everywhere) and $f'(x)$ belongs to $L^2[0, 1]$. In addition, we impose the conditions:

$$\text{for } \mathscr{D}(T_1): f(0) = f(1) = 0;$$

$$\text{for } \mathscr{D}(T_2): f(0) = 0.$$

Both of these domains are dense in $L^2[0, 1]$. Furthermore, the operators T_i are closed.

Proof. To show that $T = T_i$ is closed, $i = 1, 2$, we reason as follows. For functions with $f(0) = 0$, the inverse of $T = d/dx$ is the bounded operator

$$T^{-1}f(x) = \int_0^x f(t)\, dt.$$

Now T^{-1}, being bounded, has a closed graph. But the graph of $T = d/dx$ is just the graph of T^{-1} with the coordinate axes (i.e. the domain and range) reversed.

Strictly speaking, the above argument is incomplete, since it leaves out the boundary conditions $f(0) = f(1) = 0$ etc. This point is handled in the following way. The integral operator T^{-1} above is actually bounded from $L^2[0, 1]$ into $C[0, 1]$ (with the uniform norm on $C[\ ]$!). For, by the Schwarz inequality, the L^2-norm of $f \geqslant L^1$-norm of f, and then in turn, the L^1-norm of $f \geqslant \left|\int_0^x f(t)\, dt\right|$. Thus convergence in L^2 of $\{Tf_n\} = \{f_n'(x)\}$ ($n \to \infty$) implies uniform convergence of the original functions f_n.

This proves that the operators T_1 and T_2 are both closed. □

The above example convinces us that it would be useful to have some general criteria for showing that various operators are closed. We give here two such criteria.

Proposition (First closure criterion). *Let $T: X \to Y$ have dense domain $\mathscr{D}(T)$, and suppose that T is one to one and maps $\mathscr{D}(T)$ onto Y. Suppose further that T^{-1} is bounded. Then T is closed.*

Proof. Since $T^{-1}: Y \to X$ is bounded, T^{-1} has a closed graph. But the graph of T is just the graph of T^{-1} with the coordinate axes reversed. □

1. Bounded Operators, Closed Unbounded Operators

This simple proposition applies in many elementary situations. We saw above how it could be used to treat d/dx. Other applications will be given in due course.

Note. We mention one obvious but useful generalization of the first closure criterion. Instead of assuming that T maps $\mathscr{D}(T)$ onto Y, it suffices to assume that T maps $\mathscr{D}(T)$ onto a closed subspace Z of Y. For then we can apply the same result, with the Banach space Z replacing Y.

We turn now to the second closure criterion. This requires a bit of preface.

Sometimes an operator which is not closed has a closed extension. That is, the operator T has a graph which is not closed; by taking the closure of its graph, we define a closed operator $\bar{T}$ which is an extension of T. (This should not be confused with the example above, where we had T_1, $T_2 = d/dx$ with varying boundary conditions. There both of the operators *were* closed, and the extension was from one closed operator to another.)

This suggests a uniform procedure for finding the closure of an arbitrary unbounded operator T. Namely, consider the graph G of T, take its closure $\bar{G}$, and let $\bar{G}$ determine a new operator $\bar{T}$. If this would work in every case then the problem of finding closures would largely vanish. Where is the difficulty?

The difficulty is that $\bar{G}$ might not be the graph of a single-valued function. In fact, that is the only difficulty. A realization of this fact leads to a very general criterion for closure. This will be our second closure criterion.

As a preface for this, we must say a word about weak topologies. By a "weak topology", in this context, we simply mean a topology which is weaker than the norm topology on the Banach space in question. The point is that an operator $T: X \to Y$ which is not continuous in terms of the norm topologies on X and Y may be continuous in terms of some weaker topologies.

Remarks. The most generally useful weak topologies are those associated with Schwartz distributions. In terms of these topologies, most of the standard operators of analysis and physics are continuous. However, we should emphasize that this book leans in no essential way on distribution theory. Distributions can be used to show that certain classical operators are closed. It is for this purpose—and for this purpose only—that we use them. Thus a knowledge of distribution theory is in no way essential for an understanding of this book.

Proposition (Second closure criterion). *Let $T: X \to Y$ be an unbounded operator with dense domain $\mathscr{D}(T)$ in X. Suppose there exist Hausdorff topologies τ_1 and τ_2 on X and Y respectively, which are weaker than the norm topologies, and such that T is continuous in terms of τ_1 and τ_2. Then T has an extension to a closed operator $\bar{T}$ from X to Y.*

Proof. Let G be the graph of T in $X \times Y$, and let $\bar{G}$ be its closure. We want to define $\bar{T}$ as the operator with graph $\bar{G}$. The problem is that $\bar{G}$ might not be the graph of a single-valued function: i.e. we might have $x_n \to 0$ in the norm of X and $Tx_n \to y \neq 0$ in Y. But under the above assumptions, this does not happen. For $x_n \to 0$ in X implies

$x_n \to 0$ in the weaker topology τ_1; similarly, $Tx_n \to y$ in Y implies $Tx_n \to y$ in the topology τ_2. If $y \neq 0$, this contradicts the continuity of T in terms of τ_1 and τ_2. □

Incidentally, when an operator T has a closed extension, it is conventional to assume that this extension has been made and to say that T is closed. We shall follow this convention.

We conclude this section with two examples which illustrate the two closure criteria.

Example. Let p, r be computable reals with $1 \leq p < r < \infty$. We recall that then $L^r[0, 1] \subsetneq L^p[0, 1]$ and $\|f\|_p \leq \|f_r\|$ for any $f \in L^r$. Thus (with $p < r$) the natural injection from $L^r[0, 1]$ to $L^p[0, 1]$ is bounded of norm 1.

Now let T be the inverse mapping $T: L^p[0, 1] \to L^r[0, 1]$ defined by $Tf = f$ for $f \in L^r$. This, of course, is not an identity mapping; it is not even bounded. Its domain $\mathscr{D}(T)$ is the proper subspace $L^r[0, 1] \subsetneq L^p[0, 1]$.

However, the injection operator T *is* closed. This follows from the first closure criterion, since T^{-1} is bounded. □

Example. Let p, r be computable reals with $1 \leq p, r < \infty$. (Here we could have $p = r$.) Let T be the Fourier transform operator from $L^p(-\infty, \infty)$ to $L^r(-\infty, \infty)$. This is defined formally by

$$(Tf)(t) = \int_{-\infty}^{\infty} e^{-itx} f(x)\, dx.$$

Now, in general, neither T nor T^{-1} is bounded from L^p to L^r or vice-versa. Furthermore, a correct definition of T requires weak topologies. For the above integral makes sense only if f is integrable, i.e. if $f \in L^1$—and we have not assumed that $p = 1$.

The way out of this dilemma is to use a suitable weak topology—e.g. the topology of tempered distributions—in which the Fourier transform is well defined and continuous.

We observe, however, that it is not necessary to be fluent in the theory of tempered distributions, in order to grasp the essential features of this example. It is enough to know that such a theory exists, that it gives a well defined meaning to Tf for any $f \in L^p(-\infty, \infty)$, and that it allows us to answer the question: Does Tf belong to $L^r(-\infty, \infty)$?

We now define the domain $\mathscr{D}(T)$ to be the set of functions $f \in L^p(-\infty, \infty)$ for which $Tf \in L^r(-\infty, \infty)$. On this domain—generally a proper subset of $L^p(-\infty, \infty)$—T gives a well defined mapping into $L^r(-\infty, \infty)$.

The fact that T is closed follows immediately from the second closure criterion. □

2. The First Main Theorem

The preceding section had nothing per-se to do with computability. The notions of a closed or bounded operator are standard concepts of functional analysis. However, as we shall see, the question of whether or not a closed linear operator T is bounded largely determines the behavior of T with respect to computability.

The notion of a computability structure on a Banach space was developed axiomatically in the last chapter. We recall that there are three axioms: Linear Forms, Limits, and Norms. Two immediate consequences of these axioms are also essential: the Composition Property, which allows us to pass to a computable subsequence $\{x_{a(n)}\}$ of a computable sequence $\{x_n\}$, and the Insertion Property, which allows us to combine two computable sequences $\{x_n\}$ and $\{y_n\}$ into one.

Let X be a Banach space with a computability structure. We recall that an *effective generating set* for X is a computable sequence $\{e_n\}$ whose linear span is dense in X. In this definition, there is no requirement of "effective density"—merely that the linear span of $\{e_n\}$ be dense. However, the Effective Density Lemma from Section 5 of Chapter 2 gives:

A sequence of vectors $\{x_n\}$ is computable in X if and only if there is a computable double sequence $\{p_{nk}\}$ of (rational/complex rational) linear combinations of the e_n such that $\|p_{nk} - x_n\| \to 0$ as $k \to \infty$, effectively in k and n. Thus we do, in fact, have effective density for computable sequences $\{x_n\}$.

The following is the key theorem of this chapter.

First Main Theorem. *Let X and Y be Banach spaces with computability structures. Let $\{e_n\}$ be a computable sequence in X whose linear span is dense in X (i.e. an effective generating set). Let $T: X \to Y$ be a closed linear operator whose domain $\mathcal{D}(T)$ contains $\{e_n\}$ and such that the sequence $\{Te_n\}$ is computable in Y. Then T maps every computable element of its domain onto a computable element of Y if and only if T is bounded.*

Complement. *Under the same assumptions, if T is bounded then more can be said. The domain of T coincides with X, and T maps every computable sequence in X into a computable sequence in Y.*

Remark. We have not assumed that $\{Te_n\}$ is an effective generating set for Y.

Proof. We first assume that T is bounded and prove the stronger statement given in the Complement. Let $\{x_n\}$ be computable in X. By the Effective Density Lemma (recalled at the beginning of this section), there is a computable double sequence $p_{nk} \in X$, consisting of computable linear combinations of the e_n, such that $\|p_{nk} - x_n\| \to 0$ as $k \to \infty$, effectively in k and n.

Now $\{Te_n\}$ is computable by hypothesis, and hence by the Linear Forms Axiom, $\{Tp_{nk}\}$ is computable. Since $p_{nk} \to x_n$ as $k \to \infty$, effectively in k and n, and since T is bounded, $Tp_{n_k} \to Tx_n$ as $k \to \infty$, effectively in both variables. Hence by the Limit Axiom, $\{Tx_n\}$ is computable, as desired.

Now we come to the case where T is not bounded. We need to find a computable element $x \in \mathcal{D}(T)$ such that Tx is not computable in Y. The proof is based on two lemmas. The first of these asserts, roughly speaking, that T is "effectively unbounded". The second lemma contains the key idea of the proof.

Lemma 1. *Take the assumptions of the theorem, with T unbounded. Then there exists a computable sequence $\{p_n\}$ of finite linear combinations of the e_n such that $\{Tp_n\}$ is computable in Y and*

$$\|Tp_n\| > 10^n \|p_n\| \quad \text{for all } n.$$

Proof of Lemma 1. By hypothesis, the linear span of $\{e_n\}$ is dense in X. Since the operator T is closed, T cannot be bounded on the span of $\{e_n\}$; else T would be bounded on X. Now we sweep out the set of all finite (rational/complex rational) linear combinations of the e_n; this is easily done in an effective way by using any of the standard recursive enumerations of all finite sequences of integers. The result, by the Linear Form Axiom, is a computable sequence $\{p'_n\}$ in X which runs through all finite (rational/complex rational) linear combinations of the e_n.

By hypothesis, the sequence $\{Te_n\}$ is computable in Y. Since Tp'_n can be computed from the Te_n via linearity, it again follows from the Linear Forms Axiom that $\{Tp'_n\}$ is computable in Y.

We now construct a computable subsequence $\{p_n\}$ of $\{p'_n\}$ such that $\|Tp_n\| > 10^n\|p_n\|$ for all n. By the Composition Property, any recursive process for selecting a subsequence of indices automatically generates computable subsequences $\{p_n\}$ and $\{Tp_n\}$ in the Banach spaces X and Y respectively.

Since T is unbounded on the linear span of $\{e_n\}$, the set of ratios $\{\|Tp'_n\|/\|p'_n\|, p'_n \neq 0\}$ is unbounded. By the Norm Axiom, the sequences $\{\|Tp'_n\|\}$ and $\{\|p'_n\|\}$ are computable. So we can effectively select the desired subsequence $\{p_n\}$ of $\{p'_n\}$, with $\|Tp_n\| > 10^n\|p_n\|$, merely by waiting, for each n, until a suitable p'_n turns up. □

Lemma 2. *Let $r > 2$ be a computable real. Let $\{z_n\}$ be a computable sequence in Y with $\|z_n\| = 1$ for all n. Let $a: \mathbb{N} \to \mathbb{N}$ be a one to one recursive function which enumerates a set $A \subseteq \mathbb{N}$. (Thus the set A is recursively enumerable; it may or may not be recursive.) Then the element*

$$y = \sum_{k=0}^{\infty} r^{-a(k)} z_k$$

is computable in Y if and only if the set A is recursive.

Proof of Lemma 2. The "if" part is trivial. If A is recursive, then the series converges effectively and the Limit Axiom implies that y is computable.

For the "only if" part: We will assume that y is computable and deduce that A is recursive. Let y_n denote the n-th partial sum of the above series. Then $\{y_n\}$ is

2. The First Main Theorem

computable (Linear Forms Axiom), and y is computable (by assumption); hence $\{y - y_n\}$ is computable. By the Norm Axiom, the sequence of norms $\{\|y - y_n\|\}$ is also computable.

Now $\|y - y_n\| \to 0$, but the convergence is not monotone, and we cannot immediately deduce that the convergence is effective. We shall use the fact that $r > 2$ to establish a decision procedure for the set A (and incidentally show that $\|y - y_n\| \to 0$ effectively).

Since $r > 2$, each term in the series $\sum_{a=0}^{\infty} r^{-a}$ is strictly larger than the sum of all the following terms; the a-th term is r^{-a}, whereas $\sum_{b=a+1}^{\infty} r^{-b} = r^{-a}/(r - 1)$.

Here is the decision procedure for the set A. To test whether an integer a belongs to A, we wait until we find an n for which

$$\|y - y_n\| < r^{-a} - [r^{-a}/(r - 1)].$$

If a has occurred as some value $a(k)$, $0 \leq k \leq n$, then $a \in A$; otherwise $a \notin A$.

To justify this test, we will show that, for $k > n$, $a(k)$ takes no value $c \leq a$ (and hence in particular does not take the value a). Suppose otherwise. Let $c = a(m)$ be the *least* value taken by $a(k)$, $k > n$. Then r^{-c} exceeds the sum of all other terms $r^{-a(k)}$, $k > n$, by at least $r^{-c}[1 - (r - 1)^{-1}] \geq r^{-a}[1 - (r - 1)^{-1}]$. Now consider the series

$$y - y_n = \sum_{k>n} r^{-a(k)} z_k.$$

The term $r^{-c} z_m$ has norm r^{-c} (since $\|z_m\| = 1$). However, by the triangle inequality, the sum of the other terms has norm

$$\leq \sum_{a>c} r^{-a} = r^{-c}/(r - 1).$$

Hence

$$\|y - y_n\| \geq r^{-c} - [r^{-c}/(r - 1)],$$

and since $c \leq a$, this contradicts the defining inequality for n given in the decision procedure above. □

Proof of theorem, concluded. Following Lemma 1, we take a computable sequence $\{p_n\}$ in X such that $\{Tp_n\}$ is computable in Y and $\|Tp_n\| > 10^n \|p_n\|$. By the Norm Axiom, $\{\|Tp_n\|\}$ is computable, and we define the computable sequences

$$z_n = Tp_n/\|Tp_n\| \quad \text{in } Y,$$
$$u_n = p_n/\|Tp_n\| \quad \text{in } X.$$

Then $z_n = Tu_n$, $\|u_n\| < 10^{-n}$, and $\|z_n\| = 1$.

Let $a: \mathbb{N} \to \mathbb{N}$ be a one to one recursive function which enumerates a recursively enumerable non recursive set A. Let

$$x = \sum_{k=0}^{\infty} 10^{-a(k)} u_k,$$

$$y = \sum_{k=0}^{\infty} 10^{-a(k)} z_k.$$

Then by Lemma 2, the element y is not computable in Y. On the other hand, since $\|u_k\| < 10^{-k}$, the series for x converges effectively, so that by the Limit Axiom x is computable in X.

Finally, $Tu_n = z_n$, and the series for x and y converge (although not necessarily effectively). Since T is a closed operator, it follows that x belongs to the domain of T, and $Tx = y$. □

3. Simple Applications to Real Analysis

Before proceeding to these applications, a few comments seem in order. In the First Main Theorem above, there is only one recursion theoretic assumption on the operator T: that T map the effective generating set $\{e_n\}$ onto a computable sequence $\{Te_n\}$ in Y. All of the other assumptions on T—that it be closed, bounded or unbounded—are within the domain of standard analysis and have nothing per-se to do with computability.

As for the one recursion theoretic assumption—that T acts effectively on $\{e_n\}$—this is trivial to verify in most cases. For the effective generating sets $\{e_n\}$ are usually chosen to consist of very simple functions: e.g. the monomials $\{x^n\}$ or the trigonometric functions $\{e^{inx}\}$. For these functions, the computability of $\{Te_n\}$ is often so obvious that it seems slightly pedantic to mention it. (Example: let $T = d/dx$ and $\{e_n\} = \{x^n\}$; then we must "verify" that the sequence $\{(d/dx)x^n\}$ is computable.)

The following applications illustrate these points.

A standing convention. In each of the theorems below, "computability" will be in the standard intrinsic sense for the Banach space considered. Thus e.g. computability in $C[a, b]$ means the classical Grzegorczyk/Lacombe computability defined in Chapter 0. Henceforth we shall refer to this as "Chapter-0-computability". Computability in $L^p[a, b]$, $1 \leq p < \infty$, means the intrinsic L^p-computability defined in Chapter 2. Similarly for l^p, l_0^{∞}, $C_0(\mathbb{R})$ and other spaces—see Chapter 2 for details. Of course, a, b, and p are computable reals.

Theorem 1 (Integrals and Derivatives). (a) *The indefinite integral of a computable function $f \in C[a, b]$ is computable.*

(b) *There exist computable functions in $C[a, b]$ which have continuous derivatives, but whose derivatives are not computable.*

3. Simple Applications to Real Analysis

[The result (b) is due to Myhill [1971]. It was proved by an explicit construction in Chapter 1.]

Proof. In the First Main Theorem, we let $X = Y = C[a, b]$. As an effective generating set for X, we take the sequence of monomials $\{e_n\} = \{x^n\}$. For parts (a) and (b), we let T be the indefinite integral/derivative operator, respectively. In case (a) T is bounded; in case (b) it is not. Both operators are closed. The indefinite integral, being bounded, is automatically closed. For $T = d/dx$, we obtain a closed operator if we take the domain $\mathscr{D}(T) = C^1[a, b]$: this follows easily from either of the Closure Criteria in Section 1 above. Finally, in both cases (a) and (b), the verification of computability for $\{T(x^n)\}$ is trivial. □

Remark. In Chapter 1 we also proved that if f is computable in $C[a, b]$ and if $f \in C^2[a, b]$, then f' is computable. This involves no contradiction, even though we have an unbounded operator $T = d/dx$ which maps computable functions onto computable functions. For the operator d/dx, when restricted to $C^2[a, b]$, is not closed.

Our next example is more recondite. Although the theorem applies to $C[0, 2\pi]$, its proof requires the use of a more complicated Banach space. This theorem was proved by classical methods in Pour-El, Richards [1983a]. That classical construction was distinctly more complicated than the entire proof of the First Main Theorem.

Theorem 2 (Convergence of Fourier series). *There exists a computable function $f \in C[0, 2\pi]$ with $f(0) = f(2\pi)$ which has the following properties. The Fourier series of f is computable (i.e. the sequence of partial sums is computable), the Fourier series converges uniformly, but the series is not effectively uniformly convergent.*

Proof. Again with reference to the First Main Theorem: For X we take the space $C[0, 2\pi]$ with the points 0 and 2π identified, so that our functions must satisfy $f(0) = f(2\pi)$. We use the standard notion of Chapter-0-computability for these functions. As an effective generating set $\{e_n\}$ we take the sequence $\{e^{inx}\}$, $n = 0, \pm 1, \pm 2, \ldots$.

For Y, we take the Banach space of uniformly convergent sequences of continuous functions on $[0, 2\pi]$:

$$\{s_0(x), s_1(x), s_2(x), \ldots\}.$$

(Again, of course, we require $s_k(0) = s_k(2\pi)$.) The norm on Y is the obvious sup norm:

$$\|\{s_k(x)\}\| = \sup_{k, x} |s_k(x)|.$$

Now we must put a computability structure on Y. We say that an element $\{s_k\}$ in Y is *computable* if the sequence $\{s_k(x)\}$ is Chapter-0-computable and is effectively uniformly convergent. Similarly, a sequence $\{y_n\}$ in Y (which is a double sequence

$\{s_{nk}\}$ of functions) is *computable* if $\{s_{nk}(x)\}$ is Chapter-0-computable and, as $k \to \infty$, $s_{nk}(x)$ converges uniformly in x and effectively in k and n.

The axioms for a computability structure are easily verified for Y. Only the Norm Axiom requires a moment's thought: it holds because the double sequence $\{s_{nk}\}$ is effectively uniformly convergent as $k \to \infty$.

It would be fairly easy, although a trifle cumbersome, to describe an effective generating set for Y. We have no need of it. For, in the First Main Theorem, an effective generating set is needed in the domain X but not in the range Y.

Now that we have our range space Y, it remains only to define an appropriate operator T and verify its properties. For T, of course, we take the operator which maps each function $f \in C[0, 2\pi]$ onto its Fourier series. More precisely,

$$Tf = \{s_k\}, \qquad \text{where } s_k(x) = \sum_{m=-k}^{k} c_m e^{imx},$$

and c_m is the m-th Fourier coefficient of f.

It is well known that T is not everywhere defined (there exist continuous functions whose Fourier series do not converge); hence T is an unbounded operator. One readily verifies, however, that T is closed. Finally, as usual, the fact that $\{Te_n\}$ is computable (i.e. that we can compute the Fourier series for the functions $\{e^{inx}\}$) is trivial almost to the point of being a joke.

So now, by the First Main Theorem, there exists a computable function $f \in C[0, 2\pi]$, with $f \in \mathscr{D}(T)$, whose Fourier series $Tf = \{s_k\}$ is not computable in Y. What does this imply? Firstly, since f is in the domain of T, the Fourier series of f is uniformly convergent. Secondly since Tf is not computable in Y, we have either:

i) the sequence of partial sums $\{s_k(x)\}$ is not Chapter-0-computable, or
ii) the sequence $\{s_k(x)\}$ is not effectively uniformly convergent.

Now (i) is false: the standard method for computing Fourier coefficients gives an effective method for computing $\{s_k\}$. So we must have (ii) which proves the theorem! □

We mention in passing Fejer's Theorem, which asserts that, for any $f \in C[0, 2\pi]$, the averages

$$\sigma_k = (s_0 + s_1 + \cdots + s_k)/(k+1)$$

converge uniformly to f. Since this holds for *all* $f \in C[0, 2\pi]$, we know that the corresponding operator (from f to $\{\sigma_k\}$) must be bounded. Thus here one obtains a result in the opposite sense to that in Theorem 2:

Effective Fejer Theorem. *If $f \in C[0, 2\pi]$ is computable, with $f(0) = f(2\pi)$, then the sequence $\{\sigma_k(x)\}$ is effectively uniformly convergent to $f(x)$.*

(For more details, see Pour-El, Richards [1983a].)

3. Simple Applications to Real Analysis

L^p and l^p spaces

We turn now to the intrinsic L^p and l^p computability theories defined in Chapter 2. Our first result deals with the question: If a function f belongs to both $L^p[a, b]$ and $L^r[a, b]$, and if f is computable in L^p, is f necessarily computable in L^r? The answer, depending on p and r, is sometimes yes, sometimes no. The proof involves the First Main Theorem applied to the injection operator $Tf = f$ from L^p to L^r.

For convenience, we combine the corresponding results for $L^p[a, b]$, $L^p(\mathbb{R})$, and l^p. We remark that these results also hold for $p = \infty$ or $r = \infty$, where the corresponding spaces—as defined in Chapter 2—are:

$L^p[a, b]$: for $p = \infty$ the corresponding space is $C[a, b]$.
$L^p(\mathbb{R})$: for $p = \infty$ the corresponding space is $C_0(\mathbb{R})$.
l^p: for $p = \infty$ the corresponding space is l_0^∞.

Throughout this discussion, when p is finite, we assume that p is a computable real, $p \geq 1$. The case $p = \infty$ is as explained above. For $L^p[a, b]$, we assume the endpoints a, b are computable reals.

Theorem 3 (L^p-Computability for varying p). (a) *If $r \leq p$, then every function computable in $L^p[a, b]$ is also computable in $L^r[a, b]$. If $r > p$, then there exists a function f computable in $L^p[a, b]$, such that f belongs to $L^r[a, b]$ but is not computable in $L^r[a, b]$.*

(b) If $r \neq p$, then there exists a function f belonging to both $L^p(\mathbb{R})$ and $L^r(\mathbb{R})$, such that f is computable in $L^p(\mathbb{R})$ but not in $L^r(\mathbb{R})$.

(c) If $p \leq r$, then every sequence computable in l^p is computable in l^r. If $p > r$, then there exist sequences which are computable in l^p, belong to l^r, but are not computable in l^r.

The proof will follow. First we give some specific cases which are of independent interest.

Let us focus our attention on part (a) above, and on the case where $r = \infty$ and p is finite. Thus we are comparing computability in $L^p[a, b]$ to that in $C[a, b]$. Now $C[a, b]$ is a proper subset of $L^p[a, b]$, and it is trivial to verify that there exist computable (i.e. L^p-computable) functions in $L^p[a, b]$ which do not belong to $C[a, b]$. (Example: any discontinuous function which is L^p-computable, such as a step function with computable values and jump-points.) The following is more subtle.

Example 1. There is a continuous function which is computable in $L^p[a, b]$ but not in $C[a, b]$—i.e. a continuous function which is L^p-computable but not Chapter-0-computable.

Now, since the L^p-topology is so weak, it may not seem surprising that there are L^p-computable functions which are not Chapter-0-computable. If so, the following example may seem more striking. It is based on part (b) above with $p = \infty$ and r finite.

Example 2. There exists a function $f \in C_0(\mathbb{R}) \cap L^r(\mathbb{R})$ which is computable in C_0 but not in L^r. Spelling this out: There exists a Chapter-0-computable function f on $\mathbb{R}$, such that $f(x) \to 0$ effectively as $|x| \to \infty$, and such that $f \in L^r(\mathbb{R})$, but such that f is not computable in $L^r(\mathbb{R})$.

Proof of Theorem 3. To avoid repeating ourselves three times, we shall use the term "L^p" as a generic expression for either $L^p[a, b]$, $L^p(\mathbb{R})$, or l^p. Similarly for "L^r".

Let T be the injection operator $Tf = f$ from L^p to L^r. What is the domain of T? Since $T: L^p \to L^r$ and $Tf = f$, we need $f \in L^p$ and $f \in L^r$: thus the domain $\mathscr{D}(T) = L^p \cap L^r$.

We pause now to verify that T is closed. Since this is our first theorem dealing with L^p-spaces, we shall proceed with some care. This also provides a good opportunity to illustrate the closure criteria developed at the end of Section 1.

Begin with $L^p[a, b]$ and $L^p(\mathbb{R})$. We use the Second Closure Criterion, applied to a topology which is weaker than any of the L^p-topologies—e.g. the distribution topology. Obviously the operator $Tf = f$ is continuous in this weak topology.

For l^p, instead of distributions, we use the topology of pointwise convergence (viewing sequences in l^p as functions defined on the natural numbers).

Remark. For $L^p[a, b]$ and l^p, it turns out that T^{-1} is bounded whenever T is not. Thus we could also use the First Closure Criterion. However, this would not work for $L^p(\mathbb{R})$.

Now that we know that T is closed, we also know that T is bounded if and only if its domain $\mathscr{D}(T) = L^p$ (Closed Graph Theorem). Since, as we have seen, $\mathscr{D}(T) = L^p \cap L^r$, we conclude that T is bounded if and only if $L^p \subseteq L^r$. When does this happen? One readily verifies that:

$L^p[a, b] \subseteq L^r[a, b]$ if and only if $r \leq p$.
$L^p(\mathbb{R}) \subseteq L^r(\mathbb{R})$ if and only if $r = p$.
$l^p \subseteq l^r$ if and only if $p \leq r$.

These, the necessary and sufficient conditions for T to be bounded, are precisely the conditions given in Theorem 3—parts (a), (b), and (c), respectively—for T to preserve computability. Thus we now have almost everything we need to apply the First Main Theorem: T is closed, and we know the conditions on p and r under which T is bounded. There is one last point. We must show that T maps the effective generating set $\{e_n\}$ (for $L^p[a, b]$ or $L^p(\mathbb{R})$ or l^p as the case may be) onto a computable sequence. Well, since $Tf = f$, T maps the effective generating set onto itself. □

4. Further Applications to Real Analysis

Here we present several topics which are a bit more sophisticated than those given in Section 3. Our first and main result (Theorem 4) delineates the computability relations which hold between functions and their Fourier series and transforms.

4. Further Applications to Real Analysis

This theorem is a prototype for many similar theorems. Indeed, in a virtually identical manner, we could deal with any of the standard linear transforms of real analysis (e.g. the Hilbert transform). We mention in passing that Theorem 4 leads to an effective Plancherel Theorem and an effective Riemann-Lebesgue Lemma (Corollaries 4d and 4e).

Our second result concerns an alternative characterization of L^p-computability in terms of the Fourier coefficients and the norm. For $p = 2$, this result is a corollary of the effective Plancherel Theorem mentioned above.

Our third result also involves the properties of L^p-computability, but in a different setting. We focus on whether taking the effective L^p-closure of the "starting functions" (step functions, continuous functions, etc.) adds to the collection of computable starting functions. For continuous functions the answer is "yes", as we saw in Theorem 3 of the last section. For step functions the answer is "no", as we shall show in Theorem 5 below.

Fourier series and transforms

As a consequence of the First Main Theorem, we can easily determine the cases in which the process of taking Fourier series, Fourier coefficients, or Fourier transforms preserves computability, and also tell when it does not. First we fix some notation. For Fourier series we shall use the interval $[0, 2\pi]$, and thus define the Fourier coefficients c_k of a function $f(x)$ by:

$$c_k = (2\pi)^{-1} \int_0^{2\pi} e^{-ikx} f(x)\, dx.$$

We consider the (possibly unbounded) operator T which maps functions $f \in L^p[0, 2\pi]$ onto their sequence of Fourier coefficients $\{c_k\} \in l^r$; the domain $\mathscr{D}(T)$ is the set of such f for which $\{c_k\}$ has finite l^r-norm. Similarly we define the inverse operator T^{-1} mapping coefficient sequences $\{c_k\}$ in l^p to the corresponding functions $f(x) = \sum_{k=-\infty}^{\infty} c_k e^{ikx}$ in $L^r[0, 2\pi]$. (Note that we keep the p-norm on the domain, the r-norm on the range.) For the Fourier transform, which we denote by FT, we use:

$$FT(f)(t) = (2\pi)^{-1/2} \int_{-\infty}^{\infty} e^{-itx} f(x)\, dx.$$

Then, as is well known, the inverse transform is $FT(f)(-t)$. That is, if $g(t) = FT(f)(t)$, then $f(t) = FT(g)(-t)$.

Of course we assume that p, r are computable reals. The values $p = \infty$ and $r = \infty$ are also allowed, with the interpretation given in the preceding subsection on L^p-spaces (and also spelled out in Chapter 2). Here it is convenient to let q denote the dual variable to p, given by

$$\frac{1}{p} + \frac{1}{q} = 1.$$

Theorem 4 (Fourier series and transforms). (a) *Let f be computable in $L^p[0, 2\pi]$, and let $\{c_k\}$ be its sequence of Fourier coefficients. If $r \geqslant \max(q, 2)$, then $\{c_k\}$ is computable in l^r. Otherwise, there exist examples where f is computable in $L^p[0, 2\pi]$, $\{c_k\}$ belongs to l^r, but $\{c_k\}$ is not computable in l^r.*

(b) *The inverse mapping T^{-1} from sequences $\{c_k\} \in l^p$ to functions $f \in L^r[0, 2\pi]$ preserves computability (i.e. maps computable sequences onto computable functions) if $r \leqslant q$ and $p \leqslant 2$. Otherwise, there exist examples where $\{c_k\}$ is computable in l^p, f belongs to $L^r[0, 2\pi]$, but f is not computable in $L^r[0, 2\pi]$.*

(c) *The Fourier transform $FT: L^p(\mathbb{R}) \to L^r(\mathbb{R})$ preserves computability when $r = q$, $p \leqslant 2$, and maps some computable L^p function onto a noncomputable L^r function otherwise.*

Proof. To apply the First Main Theorem, we must go through the usual steps: (i) verify that T, T^{-1}, and FT are closed; (ii) determine the conditions under which they are or are not bounded, and (iii) verify that they act effectively on an effective generating set.

Begin with closedness. In the last Example in Section 1, we showed that FT is closed. The proofs for T and T^{-1} are similar (and easier).

The problem of deciding for which p and r the operators T, T^{-1}, and FT are bounded is by no means trivial; but it is a standard problem in classical analysis whose answer is well known. We list the results and then give some comments and literature references. The necessary and sufficient conditions for boundedness are:

for T: that $r \geqslant \max(q, 2)$;
for T^{-1}: that $r \leqslant q$ and $p \leqslant 2$;
for FT: that $r = q$ and $p \leqslant 2$.

Of course, these match precisely the cases—in (a), (b), and (c) respectively—under which the operators T, T^{-1}, and FT preserve computability.

Although these results are "folklore", we know of no source where all of them are collected in one place. However, they are easily obtainable from standard results. All of the cases of boundedness follow from the Riesz Convexity Theorem (cf. Dunford, Schwartz [1958]). For the unbounded cases, two key counterexamples are given in Zygmund [1959, Vol. 2, pp. 101–102], and all of the other examples can be derived by combining Zygmund's examples with standard arguments from Fourier analysis.

This brings us to the action of T, T^{-1}, or FT on the effective generating sets. First we must specify the effective generating sets themselves.

For $L^p[0, 2\pi]$, the domain of T, we use the effective generating set $\{e^{inx}\}$, $n = 0, \pm 1, \pm 2, \ldots$.

For l^p, the domain of T^{-1}, we take as our effective generating set the sequence $\{e_n\}$, $n = 0, \pm 1, \pm 2, \ldots$, consisting of "unit vectors" $e_n(m)$ (where $e_n(m) = 1$ for $m = n$, 0 for $m \neq n$).

For $L^p(\mathbb{R})$, the domain of the Fourier transform, we take as our effective generating set the continuous piecewise linear functions with compact support (and with rational [complex rational] coordinates for the vertices—cf. Chapter 2, Section 3).

4. Further Applications to Real Analysis

Routine computations show that T, T^{-1}, and FT act effectively on these effective generating sets. This completes the proof of Theorem 4. $\square$

The following two corollaries are so useful that we state them separately.

Corollary 4d (Effective Plancherel Theorem). *A function $f \in L^2[0, 2\pi]$ is computable in L^2 if and only if its sequence of Fourier coefficients $\{c_k\}$ is computable in l^2. A function $f \in L^2(\mathbb{R})$ is computable if and only if its Fourier transform is.*

Corollary 4e (Effective Riemann-Lebesgue Lemma). *If f is computable in $L^1(\mathbb{R})$, then its Fourier transform $FT(f)$ is computable in $C_0(\mathbb{R})$. In particular, $FT(f)(t) \to 0$ effectively as $|t| \to \infty$.*

Proofs. Corollary 4d is just the case $p = r = 2$ in parts (a), (b), (c) of Theorem 4. Corollary 4e is the case $p = 1$, $r = \infty$ of Theorem 4, part (c). $\square$

An alternative characterization of L^p-computability

Corollary 4d above suggests that there is another characterization of intrinsic L^p-computability, in which the Fourier coefficients play a major role. The next theorem shows that this is the case. For $1 < p < \infty$, the Fourier coefficients together with the norm provide such a characterization.

Theorem* (L^p-computability in terms of the Fourier coefficients and the norm). *Let p be a computable real, $1 < p < \infty$. Then a function $f \in L^p[0, 2\pi]$ is intrinsically L^p-computable if and only if:*
   (a) *the sequence $\{c_k\}$ of Fourier coefficients of f is computable, and*
   (b) *the L^p-norm of f is computable.*

The proof will not be given here, as it is quite intricate. The details can be found in Pour-El, Richards [1984]. We mention that the result fails for $p = 1$.

For the case $p = 2$, there is a much easier proof. Since this proof is short, we give it here.

Proof for $p = 2$. Suppose f is computable in L^2. Then by the Effective Plancherel Theorem (Corollary 4d above) the sequence of Fourier coefficients $\{c_k\}$ is computable in l^2. Furthermore, by the Norm Axiom, $\|f\|$ is computable.

Conversely, suppose that $\{c_k\}$ is a computable sequence of reals, and $\|f\|$ is computable. Then, since $\|f\| = \|\{c_k\}\|$, $\|\{c_k\}\|$ is computable. Hence $\left(\sum_{k=-\infty}^{\infty} |c_k|^2\right)^{1/2}$ is a computable real. Now the sequence $\left\{\sum_{k=-N}^{N} |c_k|^2\right\}$ is a computable sequence which converges monotonically to a computable limit. Hence (cf. Section 2 in Chapter 0) the convergence is effective. Thus we conclude that the sequence of partial sums $\sum_{k=-N}^{N} c_k e^{ikx}$ is a computable sequence which converges effectively to f. Hence f is computable. $\square$

Well understood functions

Recall that, for $p < \infty$, the set of L^p-computable functions is the effective closure in L^p-norm of various classes of "well understood" functions (e.g. continuous functions/step functions/piecewise linear functions, etc.). Here we ask a different question: what does the process of taking the effective L^p-closure do to the well understood functions themselves?

Let us begin with step functions—the most natural starting point for measure theoretic applications. We ask: does the process of taking the effective L^p-closure produce any new "computable" step functions? That is, are there any step functions which are L^p-computable, even though they are not computable in the "elementary" sense of having computable values and jump points? The answer turns out to be "no" (Theorem 5 below).

This result is not quite so obvious as might be supposed. To illustrate the point, suppose we ask the same question for continuous functions. That is, do there exist continuous functions which are L^p-computable, although they are not computable in the "elementary" sense of Chapter 0? The answer is "yes": cf. Example 1 following Theorem 3 in the last section.

Where does the difference lie? Presumably it is connected with the fact that step functions are determined by a finite set of real parameters, whereas continuous functions are not. Further evidence for this is obtained if we consider sequences of step functions.

Definition. A sequence $\{s_n(x)\}$ of step functions on $[a, b]$ is *computable in the elementary sense* if

$$s_n(x) = c_{ni} \quad \text{for } a_{n, i-1} \leq x < a_{ni},$$

where $a = a_{n0} < a_{n1} < \cdots < a_{nm} = b$, the number of "steps" $m = m(n)$ is a recursive function of n, $\{a_{ni}\}$ is a computable double sequence of real numbers, and $\{c_{ni}\}$ is a computable double sequence of real/complex numbers.

As previously, we ask: does the process of taking the effective L^p-closure produce any new computable sequences of step functions? The answer turns out to be "yes".

To summarize, we now have three cases: 1) step functions, 2) continuous functions 3) sequences of step functions, where we ask: does taking the effective L^p-closure increase the stock of computable starting functions? As we have seen, the answers are 1) no, 2) yes, and 3) yes.

We shall now prove these results. As already noted, 2) follows from Theorem 3. This leaves 1) (step functions) and 3) (sequences of step functions) to be dealt with. We will begin with 3).

Note. For piecewise linear functions and sequences of piecewise linear functions, the results are the same as for 1) and 3) respectively. These results could be established by the same methods.

Example. We now show that a sequence of step functions can be L^p-computable without being computable in the elementary sense.

4. Further Applications to Real Analysis

Proof. Let $a: \mathbb{N} \to \mathbb{N}$ be a one to one recursive function generating a recursively enumerable non recursive set A. For convenience, we assume $0 \notin A$, and also, in studying $a(k)$, we ignore the value $k = 0$.

Operating, for the moment, without reference to computability, suppose that $n \in A$, $n = a(k)$ with $k > 0$. Then we define a corresponding step function σ_{nk} on $[0, 1]$ by

$$\sigma_{nk}(x) = \begin{cases} 1 & \text{if } \dfrac{j}{k} \leq x \leq \dfrac{j}{k} + \dfrac{1}{k^2} \text{ for some } j, 0 \leq j < k, \\ 0 & \text{otherwise.} \end{cases}$$

Thus σ_{nk} consists of k "steps", each of height 1 and width $1/k^2$. The measure of the set $\{x: \sigma_{nk}(x) = 1\}$ is $1/k$. Hence the L^p-norm of σ_{nk} is $(1/k)^{1/p}$.

Returning now to an effective presentation, we define the double sequence of step functions s_{nk} effectively by:

$$s_{nk} = \begin{cases} \sigma_{nj} & \text{if } n = a(j) \text{ for some } j \leq k, \\ 0 & \text{otherwise.} \end{cases}$$

Then as $k \to \infty$, the computable double sequence $\{s_{nk}\}$ converges effectively in L^p-norm to the sequence $\{s_n\}$ given by:

$$s_n = \begin{cases} \sigma_{nj} & \text{if } n = a(j) \text{ for some } j, \\ 0 & \text{otherwise.} \end{cases}$$

To see that the convergence is effective, we argue as follows: If $s_{nk} \neq s_n$, then $s_{nk} = 0$, whereas $s_n = \sigma_{nj}$ for some $j > k$. Thus the L^p-norm $\|s_{nk} - s_n\| = \|\sigma_{nj}\|$, and as we saw above, $\|\sigma_{nj}\| = (1/j)^{1/p} < (1/k)^{1/p}$. Here, of course, p is fixed, and $(1/k)^{1/p} \to 0$ effectively as $k \to \infty$.

Since $\{s_n\}$ is the effective limit in L^p-norm of the computable double sequence $\{s_{nk}\}$, the sequence $\{s_n\}$ is L^p-computable.

Yet $\{s_n\}$ is not computable in the elementary sense defined above. For, if $\{s_n\}$ were computable in the elementary sense, then by definition we would have formulas for the s_n which would give a decision procedure for A. Since A is not recursive, this is impossible. In addition, the "number of steps" in the n-th step function s_n grows more rapidly than any recursive function of n. For, if $n = a(k)$, the "number of steps" in $s_n(x) = \sigma_{nk}(x)$ is just k. Now consider the "maximum number of steps" in any $s_m(x)$, $m \leq n$: it is

$$w(n) = \max\{k: a(k) \leq n\}.$$

This is just the "waiting time" defined in the first section of Chapter 0. By the Waiting Lemma, proved there, $w(n)$ is not bounded by any recursive function. □

We turn now to the case of a single step function. We work over $L^p[a, b]$, where p, a, b are computable reals, and consider an arbitrary step function $s(x)$ defined by:

$$s(x) = c_i \quad \text{for } a_{i-1} \leq x < a_i,$$

where $a = a_0 < a_1 < \cdots < a_m = b$. We call a_i an *essential transition point* if $c_{i+1} \neq c_i$, i.e. if the function really jumps at a_i.

[Note: we have not assumed a-priori that the jump-points a_i or the values c_i are computable.]

Theorem 5 (step functions). *A step function $s(x)$ is computable in $L^p[a, b]$ for $p < \infty$ if and only if all of the values c_i and all of the essential transition points a_i are computable (i.e. if and only if $s(x)$ is computable in the elementary sense).*

Proof. The "if" part is trivial.

For the "only if" part, assume that $s(x)$ is L^p-computable. This means, by definition, that there is a sequence $\{s_k\}$ of step functions, computable in the elementary sense, which converges effectively to s in L^p-norm. We have to show that s itself is computable in the elementary sense (i.e. has computable c_i and a_i).

A direct proof along these lines is not entirely trivial. A much easier proof can be given based on the First Main Theorem.

We apply the First Main Theorem to the transformation

$$T(f)(x) = \int_a^x f(u)\, du.$$

Then T is a bounded linear operator form $L^p[a, b]$ into $C[a, b]$. Clearly T maps the effective generating set $x^0, x^1, x^2, \ldots$ onto a computable sequence. Therefore T maps any function computable in $L^p[a, b]$ onto a function computable in $C[a, b]$.

Hence, since $s(x)$ is computable in $L^p[a, b]$, $T[s(x)]$ is computable in $C[a, b]$—i.e. $T[s(x)]$ is Chapter-0-computable. Since $s(x)$ is a step function, $T[s(x)]$ is a "piecewise linear" function, whose linear portions have slopes c_i, and whose transition points are the a_i. Now the Chapter 0 definition allows the effective evaluation of computable functions at computable points. Thus we can compute the slopes c_i by evaluating $T[s(x)]$ at two rational points within the same subinterval of the partition; and once the slopes are found, we can compute the a_i by solving two linear equations in two unknowns. □

Remark. Of course, pointwise evaluation makes no sense for L^p-functions, since an L^p-function is only determined up to sets of measure zero. This limitation already exists in classical analysis, without any notions of logical "effectiveness" being required. By its very nature, *an L^p-function is known only on the average.*

The point of the above proof is to map $L^p[a, b]$ into $C[a, b]$, where pointwise evaluation makes sense.

5. Applications to Physical Theory

We now discuss, from the viewpoint of computability theory, the three well-known partial differential equations of classical physics: the wave equation, the heat equation, and Laplace's equation. A similar discussion could be given for other linear differential and integral equations which occur regularly in analysis and in physical theory.

The Wave Equation

We consider the equation:

$$\frac{\partial^2 u}{\partial x^2} + \frac{\partial^2 u}{\partial y^2} + \frac{\partial^2 u}{\partial z^2} - \frac{\partial^2 u}{\partial t^2} = 0,$$

$$u(x, y, z, 0) = f(x, y, z), \tag{1}$$

$$\frac{\partial u}{\partial t}(x, y, z, 0) = 0.$$

As solutions to this equation travel with finite velocity (here made equal to 1), it is natural to consider the wave equation on compact domains. We just make the domain large enough so that "light rays" from outside the domain cannot reach any point in the domain in the time considered. Thus we define D_1 and D_2 by:

$$D_1 = \{(x, y, z) : |x| \leqslant 1, |y| \leqslant 1, |z| \leqslant 1\},$$
$$D_2 = \{(x, y, z) : |x| \leqslant 3, |y| \leqslant 3, |z| \leqslant 3\}.$$

Then if $0 < t < 2$, the solution of the wave equation on D_1 does not depend on the initial values $u(x, y, z, 0)$ outside D_2. So we may as well assume that f has domain D_2.

Consider the definitions of computability on D_1 and D_2 given in Chapter 0. Recall that they correspond respectively to the Banach spaces of continuous functions $C(D_1)$ and $C(D_2)$ with the uniform norm. Now the monomials $x^a y^b z^c$, $a, b, c \in \mathbb{N}$ provide an effective generating set both for $C(D_1)$ and for $C(D_2)$. Thus finite linear combinations of the monomials (i.e. polynomials in three variables) are dense in $C(D_i)$ ($i = 1, 2$), and a function $f \in C(D_i)$ is computable if and only if the Weierstrass approximation can be made effective (Theorem 6, Chapter 0).

The solution of the wave equation (1) is given by Kirchhoff's formula (Petrovskii [1967]).

$$u(\vec{x}, t) = \iint_{\text{unit sphere}} [f(\vec{x} + t\vec{n}) + t(\operatorname{grad} f)(\vec{x} + t\vec{n}) \cdot \vec{n}] \, d\sigma(\vec{n}), \tag{2}$$

where $\vec{x} = (x, y, z)$, $\vec{n}$ ranges over the sphere of radius 1 in $\mathbb{R}^3$, and $d\sigma(\vec{n})$ is the area measure on this sphere normalized so that the total area is 1.

Now fix t_0 so that $0 < t_0 < 2$. Let the associated solution operator $T(t_0)$ (given by Kirchhoff's formula above) be denoted by T. Since the formula contains a "grad" term, T is an unbounded operator. Furthermore T is closed. This follows because there exist weaker topologies—e.g. that of Schwartz distributions—in which the Kirchhoff operator is continuous (cf. Section 1). Finally, we must verify that T operates effectively on the monomials $x^a y^b z^c$ in the generating set. This is obvious from Kirchhoff's formula. Hence, letting $t_0 = 1$, by the First Main Theorem, we conclude:

Theorem 6 (Wave Propagation, Uniform Norm). *Consider the wave equation with the initial conditions $u = f$, $\partial u/\partial t = 0$ at time $t = 0$. Let D_1 and D_2 be the two cubes in $\mathbb{R}^3$ given above. Then there exists a computable, continuous function $f(x, y, z)$ in $C(D_2)$ such that the solution $u(x, y, z, t)$ at time $t = 1$ is continuous, but is not a computable function in $C(D_1)$.*

We now show that the situation is different if we use the energy norm instead of the uniform norm. This difference occurs even if the wave equation is treated more generally. Recall that the wave equation (1) gives a mapping from an initial function $f(x, y, z)$ on $\mathbb{R}^3$ to the solution $u(x, y, z, t)$ on $\mathbb{R}^4$. We now replace the initial condition $\partial u/\partial t = 0$ at $t = 0$ by $\partial u/\partial t = g$. This gives two initial conditions

$$u(x, y, z, 0) = f(x, y, z)$$

$$\frac{\partial u}{\partial t}(x, y, z, 0) = g(x, y, z)$$

Thus we are led to consider two Banach spaces of functions with the energy norm: one for the range space on $\mathbb{R}^4$ and one for the domain on $\mathbb{R}^3 \times \mathbb{R}^3$.

For technical reasons, we will consider a compact time interval $-M \leqslant t \leqslant M$, where $M > 0$ is a computable real. Thus the functions in our range space are defined on $\mathbb{R}^3 \times \{-M \leqslant t \leqslant M\}$.

We consider the range space first. The energy norm is defined by

$$\|u(x, y, z, t)\| = \sup_t E(u, t),$$

where (3)

$$E(u, t)^2 = \iiint_{\mathbb{R}^3} \left[|\text{grad } u|^2 + \left(\frac{\partial u}{\partial t}\right)^2 \right] dx\, dy\, dz.$$

(As is usual, $|\text{grad } u|^2 = (\partial u/\partial x)^2 + (\partial u/\partial y)^2 + (\partial u/\partial z)^2$.)

$E(u, t)^2$ is known as the energy integral. It is well known (Hellwig [1964], p. 24) that for solutions of the wave equation, $E(u, t)$ is independent of t ("conservation

5. Applications to Physical Theory

of energy"). However, for an arbitrary function u, $E(u, t)$ will depend on t. This is the reason we take the sup over t in (3).

The Banach space consists of those functions u for which the norm $\sup_t E(u, t)$ is bounded.

Although the above Banach space with the energy norm is being considered for the first time, the notion of computability is not really new. It is closely associated with the notion of "intrinsic computability" discussed in Chapter 2. This can be seen as follows. The Banach space is of mixed type—L^2 in the partial derivatives with respect to x, y, z and L^∞ in t. To obtain the computable functions we use the procedure of Chapter 2: we take the effective closure in the energy norm of the computable functions of Chapter 0. There is a small variation, however. Since the norm $E(u, t)$ involves first derivatives, we take Chapter-0-computable functions which are computably C^1—i.e. computable together with their first derivatives. Similar remarks hold for computable sequences of computable functions.

We now turn to the domain of the wave equation. It consists of pairs of functions f, g on $\mathbb{R}^3$, where f and g are initial conditions. To restrict the energy norm to $\mathbb{R}^3$, we let $t = 0$. Then the energy norm becomes the energy integral

$$\|f, g\|^2 = \iiint_{\mathbb{R}^3} [|\operatorname{grad} f|^2 + g^2]\, dx\, dy\, dz.$$

The notion of computability on this space is an obvious modification of the notion of computability on the previous space.

In summary, we have defined "computability in energy norm" for solutions $u(x, y, z, t)$ on $\mathbb{R}^4$ and for pairs of initial conditions $f(x, y, z)$ and $g(x, y, z)$ on $\mathbb{R}^3$.

There are two additional facts which we must verify in order to apply the First Main Theorem. First we must show that there is an effective generating set for the pairs of functions f, g in the domain space, and that the wave operator acts effectively on this set. Second we must show that the wave operator is bounded.

To construct an effective generating set we proceed as follows. We begin with a C^∞ function $\varphi(x) \geqslant 0$ which is supported on $[-1, 1]$, positive on the interior of this interval and computable together with all of its derivatives. For example, let

$$\varphi(x) = \begin{cases} e^{-x^2/(1-x^2)} & \text{for } |x| < 1, \\ 0 & \text{otherwise}. \end{cases}$$

Then take the sequence of functions

$$x^a y^b z^c \cdot \varphi[(x^2 + y^2 + z^2)/d^2], \quad a, b, c, d \in \mathbb{N}, d \geqslant 1.$$

Note that linear combinations of pairs of such functions generate, by effective closure in the energy norm, the set of computable pairs (f, g). Furthermore it is easily verified that the wave operator T acts effectively on pairs of functions f, g in the effective generating set on the domain space.

To show that the wave operator is bounded we note the following. The conservation of energy principle mentioned above implies that the mapping T from the initial conditions f, g to the solution $u(x, y, z, t)$ is bounded of norm 1.

Since the hypotheses of the First Main Theorem are satisfied and the solution operator is bounded we have:

Theorem 7 (Wave Propagation, Energy Norm). *Let (f, g) be a pair of initial functions on $\mathbb{R}^3$ which is computable in terms of the energy norm. Then the corresponding solution $u(x, y, z, t)$ on $\mathbb{R}^3 \times \{-M \leq t \leq M\}$ is computable in the energy norm.*

Note. A rather more complicated treatment of the wave equation with energy norm, which however deals with the entire time-interval $-\infty < t < \infty$, is given in Pour-El, Richards [1983b].

The Heat Equation

This equation is:

$$\frac{\partial u}{\partial t} = \frac{\partial^2 u}{\partial x^2} + \frac{\partial^2 u}{\partial y^2} + \frac{\partial^2 u}{\partial z^2},$$

$$u(x, y, z, 0) = f(x, y, z).$$

Its solution is given by

$$u(x, y, z, t) = \iiint_{\mathbb{R}^3} K_t(x - x', y - y', z - z') f(x', y', z') \, dx' \, dy' \, dz',$$

where

$$K_t(x, y, z) = \left(\frac{1}{4\pi t}\right)^{3/2} e^{-(x^2 + y^2 + z^2)/4t}.$$

The appropriate Banach spaces for the heat equation are $C_0(\mathbb{R}^3)$ for the domain and $C_0(\mathbb{R}^4)$ for the range, both with the uniform norm. Recall that $f(x), x \in \mathbb{R}^n$ is computable in $C_0(\mathbb{R}^n)$ if it is computable in the sense of Chapter 0 and in addition $f(x) \to 0$ effectively as $|x| \to 0$ (see Chapter 2, Section 3).

As our effective generating set for the domain space $C_0(\mathbb{R}^3)$ we use the same family of functions as with the wave equation. It is obvious that the solution formula for the heat equation operates effectively on this sequence of functions.

Since the kernel $K_t(x, y, z)$ decays rapidly as $(x^2 + y^2 + z^2) \to \infty$, the heat operator is bounded.

Thus the hypotheses of the First Main Theorem are satisfied. Since the heat operator is bounded the following theorem is an immediate corollary of the First Main Theorem.

5. Applications to Physical Theory

Theorem 8 (Heat Equation). *Let $f(x, y, z)$ be computable in $C_0(\mathbb{R}^3)$. Then the solution $u(x, y, z, t)$ of the heat equation is computable in $C_0(\mathbb{R}^4)$.*

Laplace's Equation

Here the standard problem is to find a solution u of Laplace's equation

$$\frac{\partial^2 u}{\partial x^2} + \frac{\partial^2 u}{\partial y^2} + \frac{\partial^2 u}{\partial z^2} = 0$$

on the interior of a compact domain D, such that the values of u on the boundary ∂D coincide with a preassigned continuous function f. It is well known that solutions of Laplace's equation satisfy the "maximum principle", i.e. the maximum of $|u|$ occurs on the boundary of D. Hence the mapping T from f to u is bounded of norm one. This already suggests, by the First Main Theorem, that computable boundary data f should produce computable solutions u.

The difficulties in solving this problem are geometric, relating the solution operator $T: f \to u$ to the shape of the boundary ∂D. A detailed examination of this question would lead us into potential theory, which we regard as lying outside our scope. Consequently we shall restrict our attention to a few of the standard regions which occur frequently in applications. Our list is by no means exhaustive, and many other regions could be dealt with by the same method. We consider the following regions.

1. Rectangle with computable coordinates for the "corners".
2. Cylinder with computable parameters.
3. Ball with computable center and radius.
4. Ellipsoid with computable center and axes.

Theorem 9 (Laplace's Equation). *Let D be any one of the regions listed above. Let f be a continuous function which is computable in the sense of Chapter 0. Let u be the solution of Laplace's equation which coincides with f on the boundary of D. Then u is also computable in the sense of Chapter 0.*

Proof. We have noted that T is a bounded operator. Thus the only thing that remains is to show that T maps an effective generating set onto a computable sequence. For our effective generating set we take the monomials $\{x^a y^b z^c\}$ where a, b, c are nonnegative integers. The computation of the solutions u corresponding to $f = x^a y^b z^c$ are classical, and the resulting functions are computable, effectively in a, b, c. Theorem 9 now follows as a corollary of the First Main Theorem. □

Clearly the same argument could be applied to many other regions D besides those listed above. A more general approach can also be given, using ideas similar to those in Corollary 6c of Chapter 0 (cf. Pour-El, Richards [1983b]).

Dimensions other than 3

The results in this section have been proved for the important case of space-dimension 3. All of them extend to other dimensions. Theorems 7, 8 and 9, in which the solution operator is bounded, extend mutatis-mutandis to all space-dimensions $q \geq 1$. Theorem 6 (wave equation, uniform norm) extends to space-dimensions $q \geq 2$ but not to $q = 1$ (cf. Pour-El, Richards [1981]). This is because the solution operator for the wave equation in uniform norm is unbounded for $q \geq 2$, but bounded for $q = 1$.

Part III

The Computability Theory of Eigenvalues and Eigenvectors

Chapter 4
The Second Main Theorem, the Eigenvector Theorem, and Related Results

Introduction

In this chapter, we shall be mainly concerned with operators on Hilbert space, and especially with self-adjoint operators. We will assume that our operators are, in some natural sense, "effectively determined". (The precise definition of this term is given in Section 1.) All of the standard operators of analysis and physics are effectively determined. However, we should emphasize that when an operator is called "effectively determined", this designation applies only to the operator itself, and not to the quantities derived from it, such as its eigenvalues, eigenvectors, or spectrum.

We shall address the question: Which of the quantities associated with an "effectively determined" operator are computable? For example, are the eigenvalues computable? What about the sequence of eigenvalues? We shall see that the answer is "yes" for individual eigenvalues but "no" for the sequence of eigenvalues. More precisely, these statements hold for (bounded or unbounded) self-adjoint operators and for bounded normal operators.

Here we recall some distinctions set down in Chapter 0. When we assert, as we have done for self-adjoint operators, that the individual eigenvalues are computable but the sequence of eigenvalues need not be, we mean the following. For any fixed eigenvalue, we can program a computer to compute it. However, we might need a different program for each eigenvalue—i.e. there may be no master program which gives, for each n, the n-th eigenvalue.

One might ask whether the above results for self-adjoint operators can be extended to the case of a bounded linear operator on a Banach space. The answer is "no", even for non-normal operators on Hilbert space. There exists an effectively determined bounded non-normal operator on Hilbert space which has a non-computable eigenvalue.

Let us return to self-adjoint operators, for which, as we have seen, the individual eigenvalues are computable. We can ask the same question for the eigenvectors. The answer is quite different. Even for an effectively determined *compact* self-adjoint operator, the eigenvectors associated with the eigenvalue $\lambda = 0$ need not be computable.

Here we should dispose of a triviality. Of course, for any operator, *some* of the eigenvectors are noncomputable—as we can see simply by taking a computable eigenvector and multiplying it by a noncomputable real. When we say that the eigenvectors associated with $\lambda = 0$ are not computable, we mean that *none* of them are computable.

This contrast between the computability of eigenvalues and eigenvectors can be given a physical interpretation. In quantum mechanics the eigenvalues are closely related to quantities actually measured—e.g. to the lines in the spectrum. By contrast, the eigenvectors are associated with the underlying state of the system. Our results show that the eigenvalues ae computable, whereas the eigenvectors need not be.

Besides the eigenvalues and eigenvectors, we can ask similar questions about the spectrum. (For the definition of "eigenvalue" and "spectrum" see Section 1.) Again we are mainly concerned with the self-adjoint case. We shall show that there exists a computable sequence of real numbers which belong to the spectrum and whose closure coincides with the spectrum.

We now give a brief account of the sections in this chapter.

Section 1 contains the precise definition of "effectively determined" operator. It also gives a brief review of the notions of "self-adjointness", "spectrum" and "eigenvalues" for bounded and unbounded operators.

Section 2 contains the Second Main Theorem, together with an investigation of several of its corollaries. The Second Main Theorem incorporates all of the results mentioned above for the spectra and eigenvalues of effectively determined (bounded or unbounded) self-adjoint operators. This theorem is best-possible, as we will eventually prove by suitable examples (Chapter 5, Section 8).

The proof of the Second Main Theorem is long and complicated, and it is deferred until Chapter 5.

Section 3 deals with discontinuities in the behavior of eigenvalues. For example, arbitrarily small perturbations of a self-adjoint operator can cause eigenvalues to abruptly disappear, while other eigenvalues—in quite different locations—are being suddenly created. Such discontinuities are frequently correlated with noncomputability. However, that is not the case here. Thus Section 3 (discontinuities in the eigenvalues) provides a counterpoint to Section 2 (computability of the eigenvalues).

Section 4 gives the example, promised above, of an effectively determined bounded non-normal operator with a noncomputable eigenvalue.

Sections 5 and 6 give the Eigenvector Theorem, together with its proof. This theorem asserts, as mentioned above, that there exists an effectively determined compact self-adjoint operator such that none of the eigenvectors corresponding to $\lambda = 0$ are computable. The proof of the Eigenvector Theorem is somewhat indirect. It begins, in Section 5, with a construction based on an ad-hoc (i.e. "artificial") computability structure. (Cf. Chapter 2, Section 7.) Then in Section 6 we show how to translate this ad-hoc construction into one involving the "natural" intrinsic computability structure of $L^2[0, 1]$.

Section 7 ties up some loose ends, and also, for the first time in this chapter, deals with Banach spaces other than Hilbert space. This section contains two main results. First it gives a proof of the Effective Independence Lemma, which asserts that, from any effective generating set $\{e_n\}$, we can extract a linearly independent effective generating subset. This lemma plays a role in our proof of the Eigenvector Theorem. While the result is hardly surprising, its proof is not quite so easy as might be supposed.

Second, we address the question: Are all effectively separable computability structures on a Banach space X related via (not necessarily computable) isometries? The answer turns out to be "yes" for Hilbert space, but "no" for Banach spaces in general. The "yes" part is also a step on the way to the Eigenvector Theorem (Lemma 8, below), while the "no" part is established by a counterexample given at the end of Section 7.

1. Basic Notions for Unbounded Operators, Effectively Determined Operators

Throughout most of this chapter, the underlying Banach space will be an effectively separable Hilbert space H. The inner product of two vectors $x, y \in H$ is denoted by (x, y).

We recall from Chapter 3 the notion of a closed unbounded operator. An operator $T: H \to H$ is called *closed* if T has a closed graph. In general, the domain of T is not H, but a dense linear subspace $\mathscr{D}(T)$ of H. Thus in order to defined T we must first specify the domain $\mathscr{D}(T)$ and then describe the action of T on this domain.

We now define the adjoint T^* of T. In order to motivate what follows, it is useful to begin with the familiar case of bounded operators. Let $T: H \to H$ be a bounded linear operator on H. Then, as is well known, the *adjoint* operator T^* is defined by

$$(Tx, y) = (x, T^*y) \qquad \text{for all } x, y \in H.$$

The definition of the adjoint for unbounded T is a natural extension of this. We must define the domain $\mathscr{D}(T^*)$ and the action of T^* on this domain.

Definition (first variant). Let $T: H \to H$ be a closed operator with domain $\mathscr{D}(T)$.
  a) A vector y belongs to the domain $\mathscr{D}(T^*)$ of T^* if there exists a vector z such that:

$$(Tx, y) = (x, z) \qquad \text{for all } x \in \mathscr{D}(T).$$

  b) When such a z exists, we define T^*y to be z.

Note. Thus we have the identity $(Tx, y) = (x, T^*y)$, just as in the bounded case.

It is well known (see e.g. Riesz, Sz.-Nagy [1955]) that T^* is well-defined, closed, and that its domain $\mathscr{D}(T^*)$ is dense in H.

We have preferred the above definition of T^* because it shows clearly the connection with the familiar bounded case. However, for serious work, an equivalent definition based on graphs turns out to be more powerful. First we recall that the graph of T consists of all pairs $\langle x, Tx\rangle$ with $x \in \mathscr{D}(T)$. [Here we write $\langle\ ,\ \rangle$ for ordered pairs to avoid confusion with the inner product $(\ ,\ )$.] Now the second equivalent definition is:

Definition (second variant). Let $T: H \to H$ be a closed operator, and let G be the graph of T in $H \times H$. Let $G^\perp$ be the orthogonal complement of G in $H \times H$. Then:
   a) A vector y belongs to the domain $\mathscr{D}(T^*)$ if there exists a vector z such that the pair $\langle z, y\rangle \in G^\perp$.
   b) In this case, we define T^*y to be $-z$.

The proof of equivalence of the two preceding definitions is a routine exercise (cf. Riesz, Sz.-Nagy [1955]). The "$-$" in part (b) of the second definition is not a misprint. It occurs because of formal manipulations involving the inner product on $H \times H$.

We now come to the basic:

Definition. A closed operator $T: H \to H$ (bounded or unbounded) is said to be *self-adjoint* if it coincides with its adjoint, i.e. if $T = T^*$.

In the unbounded case, it is important to stress that two operators are considered to be "equal" if and only if they have the same graph. As we recall from Chapter 3, an unbounded closed operator T_1 may possess a proper extension T_2: this means that the domain $\mathscr{D}(T_1) \subsetneqq \mathscr{D}(T_2)$, although T_2 coincides with T_1 on their common domain of definition.

The most important operators of analysis and physics—and, in particular, the so-called "observables" of quantum mechanics—are either self-adjoint or possess self-adjoint extensions. Throughout the remainder of this book, unless stated otherwise, we shall be concerned with (bounded or unbounded) operators which are self-adjoint.

There is one minor exception. In the bounded case, we shall sometimes consider normal operators. A bounded operator $T: H \to H$ is said to be *normal* if it commutes with its adjoint, i.e. if $T^*T = TT^*$. Normal operators possess many of the same properties as self-adjoint operators. In particular, the Second Main Theorem also holds for bounded normal operators, as we shall prove in Section 6 of Chapter 5.

Eigenvalues and spectrum

The structure of a bounded or unbounded self-adjoint operator is determined to a substantial degree by the eigenvalues and spectrum of that operator.

Definition (spectrum). Let T be a closed operator (bounded or unbounded). A number λ belongs to the *spectrum* of T if the operator $(T - \lambda)$ does not have a bounded inverse.

We observe that, for λ *not* to be in spectrum (T), the inverse $(T - \lambda)^{-1}$ must be bounded, whether T itself is bounded or not.

1. Basic Notions for Unbounded Operators, Effectively Determined Operators

It is well known (cf. Riesz, Sz.-Nagy [1955]) that the spectrum is a closed set, and that when T is self-adjoint, the spectrum is real. Hence, for self-adjoint T, the inverse $(T - \lambda)^{-1}$ exists and is bounded for all $\lambda = \alpha + i\beta$ with $\beta \neq 0$; in particular, $(T - i)^{-1}$ exists. In addition, for self-adjoint T, the spectrum is a bounded set if and only if T is bounded.

Definition (eigenvalues). A number λ is called an *eigenvalue* of T if there exists a nonzero vector x (the corresponding eigenvector) such that $Tx = \lambda x$.

The eigenvalues form a subset of the spectrum, which in general is a proper subset. Points of the spectrum which are not eigenvalues are commonly said to belong to "the continuous spectrum".

The eigenvalues and continuous spectrum have a direct physical significance in quantum mechanics. For example, when the emissions from a light source are observed in the spectroscope, the eigenvalues are closely related to the appearance of "bright lines", whereas the continuous spectrum is reflected in the presence of continuous bands of light.

Effectively determined operators

Up to now, there has been no mention within this section of computability. We have simply set down some standard facts from functional analysis. Now we must define what it means for an operator to "act effectively". We shall call such operators *effectively determined*.

Recall that, by assumption, the space H is effectively separable. This means that there is a computability structure defined on H, and that H has an effective generating set $\{e_n\}$—i.e. a computable sequence $\{e_n\}$ whose linear span is dense in H. (See Chapter 2.)

Operators map vectors into vectors. Thus to "determine" an operator means to know, in some effective manner, how the operator acts on computable vectors and computable sequences of vectors. In the unbounded case, we need to know one thing more. We have seen that, to define an unbounded operator, we really need to know its graph. Thus we should expect the definition of an "effectively determined" unbounded operator to involve an effective specification of the graph.

Since the graph is contained within $H \times H$, we must first define "computability" for the cartesian product. The definition is obvious. A sequence of points $\langle x_n, y_n \rangle \in H \times H$ is called *computable* if $\{x_n\}$ and $\{y_n\}$ are computable in H.

Definition (Effectively determined operator). A closed operator $T: H \to H$ is *effectively determined* if there is a computable sequence $\{e_n\}$ in H such that the pairs $\{\langle e_n, Te_n \rangle\}$ form an effective generating set for the graph of T.

We recall that, by the definition of "effective generating set", this means that $\{\langle e_n, Te_n \rangle\}$ is computable in $H \times H$, and that the linear span of $\{\langle e_n, Te_n \rangle\}$ is dense in the graph of T. In particular, this implies that $\{e_n\}$ and $\{Te_n\}$ are computable and that the span of $\{e_n\}$ is dense in H. Thus, as a corollary of the above definition,

$\{e_n\}$ is also an effective generating set for H. The converse of this corollary is false, however. When the operator T is unbounded, the density of (linear span of $\{e_n\}$) in H does *not* imply that $\{\langle e_n, Te_n\rangle\}$ spans a dense subset of the graph.

The following may help to explain what could go wrong. We saw in Chapter 3, Section 1 that a closed operator T_1 can have a proper closed extension T_2, where $\mathscr{D}(T_1) \subsetneqq \mathscr{D}(T_2)$. Suppose we are really interested in the extended operator T_2. Suppose, however, that we choose an effective generating set $\{e_n\}$ contained within the smaller domain $\mathscr{D}(T_1)$. Now $\mathscr{D}(T_1)$ is dense in $\mathscr{D}(T_2)$, since it is dense in H. Thus $\{e_n\}$ is a perfectly good effective generating set for H. But on $H \times H$, the pairs $\langle e_n, Te_n\rangle$ span at most the graph of T_1. For they lie within the graph of T_1, and this graph is *closed*. Thus $\{\langle e_n, Te_n\rangle\}$ cannot approach those points on graph (T_2) which lie outside of graph (T_1).

Of course, these difficulties can only occur for unbounded operators. Consider, by contrast, a bounded operator T. Then T is continuous, and the convergence in H of $\{e_n\}$ (or linear combinations thereof) automatically implies convergence of the pairs $\{\langle e_n, Te_n\rangle\}$.

In summary, a bounded operator T is effectively determined if and only if it maps an effective generating set $\{e_n\}$ for H onto a computable sequence $\{Te_n\}$.

As mentioned above, all of the standard operators of analysis and physics are effectively determined. The most interesting of these are unbounded.

2. The Second Main Theorem and Some of Its Corollaries

We recall that H denotes an effectively separable Hilbert space—i.e. a Hilbert space with a computability structure for which there is an effective generating set $\{e_n\}$.

Second Main Theorem. *Let $T: H \to H$ be an effectively determined (bounded or unbounded) self-adjoint operator. Then there exists a computable sequence of real numbers $\{\lambda_n\}$ and a recursively enumerable set A of natural numbers such that*:

i) *Each $\lambda_n \in$ spectrum (T), and the spectrum of T coincides with the closure of $\{\lambda_n\}$.*

ii) *The set of eigenvalues of T coincides with the set $\{\lambda_n: n \in \mathbb{N} - A\}$. In particular, each eigenvalue of T is computable.*

iii) *Conversely every set which is the closure of $\{\lambda_n\}$ as in (i) above occurs as the spectrum of an effectively determined self-adjoint operator.*

iv) *Likewise, every set $\{\lambda_n: n \in \mathbb{N} - A\}$ as in (ii) above occurs as the set of eigenvalues of some effectively determined self-adjoint operator T. If the set $\{\lambda_n\}$ is bounded, then T can be chosen to be bounded.*

Note. Concerning the boundedness of T: In (iii), where $\{\lambda_n\}$ determines the entire spectrum, the boundedness of $\{\lambda_n\}$ implies the boundedness of T—for, as is well known (Riesz, Sz.-Nagy [1955]), the spectral norm of a self-adjoint operator coincides with its norm. However, in (iv), where $\{\lambda_n\}$ gives only the set of eigenvalues (and not the entire spectrum), the boundedness of T must be considered separately.

2. The Second Main Theorem and Some of Its Corollaries

The proof of this theorem is long and complicated. It is deferred until Chapter 5. In fact, it forms the entire content of Chapter 5. We turn now to some consequences of the Second Main Theorem.

As stated in (ii) above, the individual eigenvalues of T are computable. However, the sequence of eigenvalues need not be, as we now show.

Theorem 1 (The sequence of eigenvalues). *There exists an effectively determined bounded self-adjoint operator $T: H \to H$ whose sequence of eigenvalues is not computable.*

Proof. We use the counterexample asserted in (iv) above. (For its details, see Chapter 5, Section 8.) Let $\{\lambda_n\}$ be the computable sequence $\lambda_n = 2^{-n}$, and let A be any recursively enumerable non recursive set of integers. Then by (iv) there exists an effectively determined bounded self-adjoint operator T whose eigenvalues coincide with the set $\{\lambda_n: n \in \mathbb{N} - A\}$. But now any effective listing of these λ_n would give an effective listing of the corresponding set of n's, i.e. an effective listing of the complement of A. Since A is not recursive, this is impossible. □

For the special case of compact operators, the phenomenon exhibited in Theorem 1 cannot occur. Namely, as a consequence of (i) above, we have:

Theorem 2 (Compact operators). *Let $T: H \to H$ be an effectively determined compact self-adjoint operator. Then the set of eigenvalues of T forms a computable sequence of real numbers.*

Proof. It is well known (Riesz, Sz.-Nagy [1955]) that, if T is compact, the spectrum of T consists of isolated eigenvalues $\lambda \neq 0$ together with $\{0\}$ as their only possible limit point. Now take the sequence $\{\lambda_n\}$ given by (i). Since the eigenvalues $\lambda \neq 0$ are isolated, and $\{\lambda_n\}$ is dense in the spectrum, it follows that every eigenvalue $\lambda \neq 0$ equals λ_n for some n.

We now consider the value $\lambda = 0$, which may or may not be an eigenvalue. To deal with this, we first extract the computable subsequence $\{\lambda'_n\}$ of all $\lambda_n \neq 0$, and then include or exclude the value $\lambda = 0$ according as it is an eigenvalue or not. □

Just as the negative result (iv) gave information about the sequence of eigenvalues, the negative result (iii) gives information about the operator norm.

Theorem 3 (The operator norm). *There exists an effectively determined bounded self-adjoint operator $T: H \to H$ whose norm is not a computable real.*

Proof. We recall that the "norm" of an operator T is defined to be $\sup \{\|Tx\|/\|x\|: x \neq 0\}$. We shall also use the "spectral norm" defined as $\sup \{|\lambda|: \lambda \in \text{spectrum}(T)\}$. It is well known (Riesz, Sz.-Nagy [1955], Halmos [1951]) that for self-adjoint operators, the norm and the spectral norm coincide.

Now let $a: \mathbb{N} \to \mathbb{N}$ be a one to one recursive function listing a recursively enumerable non recursive set A. Let

$$\lambda_n = \sum_{k=0}^{n} 2^{-a(k)}.$$

Then $\{\lambda_n\}$ is a computable monotone sequence converging noneffectively to a non-computable real number α (cf. Chapter 0, Sections 1 and 2).

Let T be the operator, promised in (iii) above, whose spectrum is the closure of $\{\lambda_n\}$. Then the spectral norm of T (= the norm of T) is sup $\{\lambda_n\} = \alpha$, a noncomputable real. □

3. Creation and Destruction of Eigenvalues

It is a well-known fact that eigenvalues of a self-adjoint operator can be instantaneously created and destroyed—that is, their behavior can be highly discontinuous. Such discontinuities often lead to noncomputability. Eigenvalues furnish an exception to this rule. For, as this section will show, we have discontinuity. By contrast, the key theorem of the last section asserts that the eigenvalues are computable.

What do we mean here by discontinuity? We mean that there is a one-parameter family T_ε of bounded self-adjoint operators, such that T_ε varies continuously with ε, but the behavior of the eigenvalues is discontinuous.

What do we mean when we say that T_ε "varies continuously with ε"? We mean that we have continuity in the sense of uniform convergence on the unit ball, i.e. in the operator norm defined by $\|T\| = \sup \{\|Tx\|: \|x\| \leq 1\}$. Thus to say that $T_\varepsilon \to T$ as $\varepsilon \to 0$ means that the operator norm $\|T_\varepsilon - T\| \to 0$. This is a very strong condition: in fact the operator norm topology is the strongest of the topologies commonly employed for operators.

Theorem 4 (Creation and destruction of eigenvalues). *There exists a one parameter family $\{T_\varepsilon\}$ of bounded self-adjoint operators on a separable Hilbert space H such that:*

i) *T_ε varies continuously with ε in terms of the operator norm topology on $\{T_\varepsilon\}$.*

ii) *For $\varepsilon = 0$, the operator T_0 has the unique eigenvalue $\lambda = 0$, and this eigenvalue is of multiplicity one.*

iii) *For all $\varepsilon \neq 0$ and sufficiently close to zero, (a) T_ε has no eigenvalue near zero, (b) T_ε has an eigenvalue near each of the points ± 1, and (c) all eigenvalues are of multiplicity one.*

Proof. Let H be the direct sum of $L^2[-1, 1]$ and an element δ of norm one generating a 1-dimensional Hilbert space (δ). Thus δ is orthogonal to $L^2[-1, 1]$. We denote functions in $L^2[-1, 1]$ by the letters $f, g, h, \ldots$. We define the operator

3. Creation and Destruction of Eigenvalues

T_ε on H by:

$$T_\varepsilon[f(x)] = xf(x) + \varepsilon \cdot \int_{-1}^{1} f(x)\,dx \cdot \delta,$$

$$T_\varepsilon[\delta] = \varepsilon \cdot 1 \quad \text{(a constant function on } [-1, 1]).$$

It is easy to verify that T_ε is self-adjoint. Also, we see at once that T_ε varies continuously with ε in terms of the operator norm.

Now for $\varepsilon = 0$, T has the eigenvalue 0 with eigenvector δ.

We shall show that for all sufficiently small $\varepsilon \neq 0$, T_ε has an eigenvalue of multiplicity one located near each of the points $\lambda = \pm 1$, and no eigenvalue near 0. Thus the eigenvalue 0 is destroyed, whereas λ near ± 1 are created.

Take $\varepsilon \neq 0$. Now any eigenvector of T_ε must involve δ. Hence, multiplying by a scalar, we may assume that the eigenvector has the form $f(x) + \delta$. Let λ be the corresponding eigenvalue, so that

$$T_\varepsilon[f(x) + \delta] = \lambda[f(x) + \delta].$$

Recalling the definition of T_ε, and equating the $L^2[-1, 1]$ and (δ) components, we obtain:

$$xf(x) + \varepsilon = \lambda f(x),$$

$$\varepsilon \cdot \int_{-1}^{1} f(x)\,dx = \lambda.$$

But the first equation gives

$$f(x) = -\varepsilon/(x - \lambda),$$

and from the second equation we have

$$-\varepsilon^2 \int_{-1}^{1} \frac{dx}{x - \lambda} = \lambda.$$

We now focus our attention on the last two equations. We must determine the values of λ which satisfy these equations, and for which the associated $f(x) \in L^2[-1, 1]$. All values $\lambda \in [-1, 1]$ are ruled out, since the function $f(x) = -\varepsilon/(x - \lambda)$ is not L^2 on $[-1, 1]$. Thus we need examine only $\lambda > 1$ and $\lambda < -1$. Let us consider $\lambda > 1$; the other case is similar. For $\lambda > 1$, the function $f(x)$ is L^2 on $[-1, 1]$, and so we need only ask whether or not the above equations are satisfied. We claim that, for each $\varepsilon \neq 0$, there is a unique $\lambda > 1$ which satisfies the last displayed equation above. Furthermore, $\lambda \downarrow 1$ as $\varepsilon \to 0$. To see this, we rewrite this equation as

$$\int_{-1}^{1} \frac{dx}{x - \lambda} = \frac{-\lambda}{\varepsilon^2}.$$

Fix any $\varepsilon \neq 0$. Then as $\lambda \downarrow 1$, the integral above decreases to $-\infty$, whereas the fraction increases to $-1/\varepsilon^2$. Hence, by the intermediate value theorem, there is a unique solution λ. $\square$

4. A Non-normal Operator with a Noncomputable Eigenvalue

In the Second Main Theorem and its extensions we assert that the eigenvalues of an effectively determined self-adjoint (or normal) operator are computable. Since the proof (cf. Chapter 5) uses the spectral theorem, the result is closely tied to the assumed self-adjointness/normality of the operator. This restriction is necessary as we now show.

The following theorem is proved for the Hilbert space $H = L^2[0, 1]$ with its standard computability structure. However, as we will show later (Lemma 8, Section 6), the result could be transferred to any effectively separable Hilbert space.

Theorem 5 (Noncomputable eigenvalues). *There exists an effectively determined bounded operator $T: H \to H$ (not self-adjoint or normal) which has a noncomputable real number α as an eigenvalue.*

Proof. Let $a: \mathbb{N} \to \mathbb{N}$ be a one to one recursive function which enumerates a recursively enumerable non recursive set A. Let $H = L^2[0, 1]$, and let $\{e_n\}$ be a computable orthonormal basis for H.

To define T, it suffices to give the value of $T(e_n)$ for all n. We define, for each n:

$$T(e_n) = \left(\sum_{k=0}^{n} 4^{-a(k)}\right) e_n + 2^{-a(n)} \cdot \sum_{k=0}^{n-1} 2^{-a(k)} e_k.$$

Now we show that T is bounded. Write $T = T_1 + T_2$, where T_1 and T_2 correspond respectively to the first and second terms in the above expression for $T(e_n)$. Then T_1 is a bounded self-adjoint operator: in terms of the basis $\{e_n\}$, it corresponds to a diagonal matrix with the bounded sequence of eigenvalues $\left\{\sum_{0}^{n} 4^{-a(k)}\right\}$. For T_2 we reason as follows. The vectors $\sum_{0}^{n-1} 2^{-a(k)} e_k$ are bounded in norm by $\sum 2^{-a(k)} \leq 2$. Hence for each n, $\|T_2(e_n)\| \leq 2 \cdot 2^{-a(n)}$. Take an arbitrary vector x:

$$x = \sum_{n=0}^{\infty} c_n e_n.$$

Since $\{e_n\}$ is an orthonormal basis, the norm $\|x\|$ is just the l^2-norm of the sequence $\{c_n\}$. Now

$$\|T_2 x\| \leq \sum_{n=0}^{\infty} |c_n| \|T_2 e_n\| \leq 2 \sum_{n=0}^{\infty} |c_n| \cdot 2^{-a(n)}.$$

5. The Eigenvector Theorem

Since $\{2^{-a(n)}\}$ is an l^2-sequence, it follows from the Schwarz inequality (for sequences) that T_2 is bounded. Hence $T = T_1 + T_2$ is bounded.

Now we show that T has a noncomputable eigenvalue. Namely, we show that the vector

$$\sum_{k=0}^{\infty} 2^{-a(k)} e_k$$

is an eigenvector with the eigenvalue

$$\alpha = \sum_{k=0}^{\infty} 4^{-a(k)}.$$

First we note the identity:

$$T\left[\sum_{k=0}^{n} 2^{-a(k)} e_k\right] = \left(\sum_{k=0}^{n} 4^{-a(k)}\right) \cdot \left(\sum_{k=0}^{n} 2^{-a(k)} e_k\right).$$

[This identity follows by a straightforward induction on n, using the definition of $T(e_n)$ given above. Of course, this identity is also the motivation for our definition of $T(e_n)$.]

Letting $n \to \infty$, we deduce that

$$T\left[\sum_{k=0}^{\infty} 2^{-a(k)} e_k\right] = \left(\sum_{k=0}^{\infty} 4^{-a(k)}\right) \cdot \left(\sum_{k=0}^{\infty} 2^{-a(k)} e_k\right).$$

Finally, the eigenvalue $\alpha = \sum_{0}^{\infty} 4^{-a(k)}$ is not a computable real. □

Note. These last few steps show why it is essential that the operator T be non-normal. For the above construction involves a sequence of eigenvectors $\left\{\sum_{k=0}^{n} 2^{-a(k)} e_k\right\}$, all very close together and converging to $\sum_{0}^{\infty} 2^{-a(k)} e_k$. These eigenvectors have the slightly different eigenvalues $\sum_{k=0}^{n} 4^{-a(k)}$. With a normal operator, distinct eigenvalues would force the vectors to be orthogonal and not close together.

5. The Eigenvector Theorem

In this and the following section we will prove:

Theorem 6 (The Eigenvector Theorem). *Let $H = L^2[0, 1]$ with its intrinsic computability structure. There exists an effectively determined compact self-adjoint operator $T: H \to H$ with the following properties.*

(1) *The number $\lambda = 0$ is an eigenvalue of T of multiplicity one (i.e. the space of eigenvectors corresponding to $\lambda = 0$ is one dimensional).*
(2) *None of the eigenvectors corresponding to $\lambda = 0$ is computable.*

As noted in the Introduction to this chapter, the proof is indirect. In this section we shall prove the following weaker result.

Eigenvector Theorem (preliminary form). *There exists an ad hoc computability structure on H, and an operator $T: H \to H$ which is effectively determined in terms of this ad-hoc structure and such that T is compact, self-adjoint, and satisfies conditions (1) and (2) above. Furthermore, H is effectively separable in terms of this structure.*

In Section 6 we shall show how to translate this preliminary theorem into its desired final form (Theorem 6 above), involving the ("natural") intrinsic computability structure on $L^2[0, 1]$.

Remarks. We recall (Chapter 2, Section 7) that an *ad hoc* computability structure is a non-intrinsic structure—i.e. a structure that can be regarded as "artificial"—which nevertheless satisfies the axioms for computability on a Banach space. The proof of the Eigenvector Theorem provides the main application of ad hoc computability given in this book. This approach is by no means an exercise in "fancy" technique. A bit of explanation seems in order.

In the Eigenvector Theorem, we have two objects to deal with: (a) the operator T, and (b) the computability structure. Obviously, to build such a counterexample, at least one of these must be somewhat intricate. In the preliminary (ad hoc) construction, the operator T is very simple, and it is the computability structure which carries the essential ideas of the counterexample. In Section 6 we reverse field, showing how *any* such counterexample can be translated into one involving the natural computability structure and a complicated operator.

Of course, the final form of the theorem (as completed in Section 6) is the result we mainly want, since it is the natural computability structure on $L^2[0, 1]$ which is of primary interest. However, in this final form, the operator T becomes so complicated that its intuitive meaning is lost. It seems that the final form of the operator would be much harder to discover, ab initio, than the slightly perturbed computability structure with which we begin this construction.

Proof of the Eigenvector Theorem (*preliminary form*). We begin by defining the ad hoc computability structure on $H = L^2[0, 1]$. First we take any one of the standard computable orthonormal bases $\{e_n\}$ for $L^2[0, 1]$: e.g. let $\{e_n\}$ be the sequence of functions $\{e^{2\pi i m x}\}$, with the integers $m = 0, \pm 1, \pm 2, \ldots$ mapped onto the natural numbers $n = 0, 1, 2, \ldots$ in a standard computable way. For convenience, assume that $m = 0$ corresponds to $n = 0$.

Let H_0 be the closed subspace of H spanned by the vectors $\{e_1, e_2, \ldots\}$; thus H_0 consists of all vectors in H which are orthogonal to e_0. The vector e_0 will play a special role in our construction, and therefore for typographical clarity we write:

$$\Lambda = e_0.$$

5. The Eigenvector Theorem

[We emphasize that there is nothing "exotic" about Λ; in fact $\Lambda = e_0 = e^{2\pi i 0 x}$ is just the constant function 1.]

Now the computability structure which we shall place on H will coincide with the natural structure on the subspace H_0, but will behave in an "artificial" manner with respect to the vector Λ. In this ad hoc structure, the vector Λ will not be computable. Here are the details.

Let $a: \mathbb{N} \to \mathbb{N}$ be a recursive function which enumerates a recursively enumerable non-recursive set A in a one to one manner. We assume that $0 \notin A$. We define a sequence of positive reals $\{\alpha_n\}$ and a single positive real number γ by setting:

$$\alpha_n = 2^{-a(n)} \qquad \text{for } n \geq 1,$$

$$\gamma^2 = 1 - \sum_{n=1}^{\infty} \alpha_n^2.$$

It is important to observe that γ is not computable. For if it were, then the series $\sum \alpha_n^2$ would converge effectively, contradicting the fact that A is not recursive.

Now we define the vector

$$f = \gamma \Lambda + \sum_{n=1}^{\infty} \alpha_n e_n.$$

We observe that $\|f\| = 1$.

Remarks. In terms of the natural computability structure on $L^2[0, 1]$, the vector f is not computable (since the coefficient γ is not computable). In the ad hoc structure below, we will declare f to be "computable". We will declare, further, that the standard notion of computability holds on H_0. This essentially determines the ad hoc structure. It takes some work, however, to show that this structure satisfies the axioms for computability on a Banach space. (Cf. Lemmas 1, 2, and 3 below.)

We put an ad hoc computability structure on H by specifying that

$$\{f, e_1, e_2, e_3, \ldots\}$$

is an effective generating set.

Equivalently, a sequence $\{x_n\}$ of vectors in H is *ad hoc computable* if:

$$x_n = \beta_n f + y_n,$$

where $\{\beta_n\}$ is a computable sequence of complex numbers, and $\{y_n\}$ is a computable sequence of vectors (in the standard sense) in H_0.

As mentioned above, we will prove in Lemma 3 that this definition satisfies the axioms for a computability structure.

The operator T. We define T by setting:

$$T\Lambda = 0,$$

$$Te_n = 2^{-n} e_n \qquad \text{for } n \geq 1.$$

Since $\{\Lambda = e_0, e_1, e_2, \ldots\}$ form an orthonormal basis for H, T is self-adjoint. Since the sequence $2^{-n} \to 0$, T is compact. Clearly 0 is an eigenvalue of T of multiplicity one; the corresponding eigenvectors are the scalar multiples of Λ.

We shall show in Lemma 4 that T is effectively determined in terms of the ad hoc computability structure which we have defined on H.

We show in Lemma 5 that no nonzero multiple of the eigenvector Λ is computable in terms of the ad hoc structure.

By combining Lemmas 3, 4, and 5, we obtain all of the conditions stated in the preliminary form of the Eigenvector Theorem.

Statement and proof of Lemmas 1–5. We begin by noting that some of these lemmas are stated in the complete generality of an arbitrary Hilbert space H with a computability structure. In these cases, obviously, we must use nothing but the axoms for computability on a Banach space. When H is a specific space (e.g. $H = L^2[0, 1]$) we shall say so.

The following is important, but hardly deserves to be called a lemma. Let H be any Hilbert space with a computability structure. If $\{x_n\}$ and $\{y_n\}$ are computable sequences of vectors in H, then the double sequence of inner products $\{(x_n, y_m)\}$ is computable. For

$$(x_n, y_m) = \tfrac{1}{4}[\|x_n + y_m\|^2 - \|x_n - y_m\|^2 + i\|x_n + iy_m\|^2 - i\|x_n - iy_m\|^2].$$

The Linear Forms Axiom gives us the computability in H of the double sequences $\{x_n \pm y_m\}$ and $\{x_n \pm iy_m\}$. Then the Norm Axiom implies that $\{(x_n, y_m)\}$ is computable. (Of course, to handle the double sequences, we use one of the standard recursive pairing functions from $\mathbb{N} \times \mathbb{N}$ onto $\mathbb{N}$.)

Lemma 1. *Let H be a Hilbert space with an arbitrary computability structure imposed upon it. Suppose there exists a computable orthonormal basis $\{e_n\}$ for H. Take any sequence of vectors $\{x_n\}$, given by*

$$x_n = \sum_{k=0}^{\infty} c_{nk} e_k.$$

Then the sequence $\{x_n\}$ is computable in H if and only if:
i) the double sequence $\{c_{nk}\}$ of "Fourier coefficients" is computable, and
ii) the series $\sum_{k=0}^{\infty} |c_{nk}|^2$ converges effectively in k and n.

Proof. The "if" part is trivial: the computability of $\{x_n\}$ follows immediately from the Linear Forms and Limit Axioms. For the "only if", take a computable sequence $\{x_n\}$. As we have seen, the sequence of inner products (x_n, e_k) is computable; this gives (i). To obtain (ii), we use the Norm Axiom. This tells us that $\|x_n\|^2 = \sum_{k=0}^{\infty} |c_{nk}|^2$ is a computable sequence of reals. Since the sequences of partial sums $\sum_{i=0}^{k} |c_{ni}|^2$ are

5. The Eigenvector Theorem

computable and converge monotonically to a computable sequence of reals, the convergence is effective in k and n. ∎

The next lemma embodies one of the key steps on the way to proving that the ad hoc structure on $L^2[0, 1]$ satisfies the axioms. The point is that the vector x considered there is *not* assumed to be computable. In terms of Lemma 1 above, x satisfies condition (i) but not necessarily condition (ii).

Lemma 2. *Let H be any Hilbert space with a computability structure and with a computable orthonormal basis $\{e_n\}$ for H. Let $\{y_m\}$ be any computable sequence of vectors in H. Take any vector $x = \sum c_k e_k$ for which the sequence of "Fourier coefficients" $\{c_k\}$ is computable (condition (i) of Lemma 1). Then the sequence of inner products $\{(x, y_m)\}$ is computable.*

Proof. We have not assumed that the series $\sum |c_k|^2$ for $\|x\|^2$ converges effectively. However, there exists an integer M which bounds $\|x\|$.

Now let the "Fourier expansions" of the y_m be

$$y_m = \sum_{k=0}^{\infty} d_{mk} e_k.$$

Since $\{y_m\}$ is computable, the double sequence $\{d_{mk}\}$ is computable, and the series $\sum_{k=0}^{\infty} |d_{mk}|^2$ converge effectively in k and m by Lemma 1. Furthermore,

$$(x, y_m) = \sum_{k=0}^{\infty} c_k \bar{d}_{mk} \qquad \text{for all } m.$$

Since the sequences $\{c_k\}$ and $\{d_{mk}\}$ are computable, to prove that $\{(x, y_m)\}$ is computable we have only to show that in the above series the convergence is effective in k and m.

Since $\sum_{k=0}^{\infty} |d_{mk}|^2$ converges effectively, there is a recursive function $e(m, N)$ such that

$$\left(\sum_{k=e(m,N)}^{\infty} |d_{mk}|^2 \right)^{1/2} \leq \frac{2^{-N}}{M} \qquad \text{for all } m, N.$$

Since the norm of x is dominated by M, the Schwarz inequality implies that

$$\sum_{k=e(m,N)}^{\infty} |c_k \bar{d}_{mk}| \leq 2^{-N} \qquad \text{for all } m, N.$$

Thus to compute $(x, y_m) = \sum_{k=0}^{\infty} c_k \bar{d}_{mk}$ to within an error of 2^{-N}, we merely compute the first $e(m, N)$ terms of the series. ∎

From Lemma 2 we derive:

Lemma 3. *The "ad hoc" computability structure defined above on H satisfies the axioms for computability on a Banach space. Furthermore, this computability structure is effectively separable.*

Proof. Only the Norm Axiom causes any difficulty. Given a computable sequence $\{x_n\}$ in H, we need to compute $\{\|x_n\|\}$. By definition:

$$x_n = \beta_n f + y_n,$$

with $\{\beta_n\}$ computable in $\mathbb{C}$, and $\{y_n\}$ computable in H_0. We recall that

$$f = \gamma \Lambda + \sum_{n=1}^{\infty} \alpha_n e_n,$$

where $\{\alpha_n\}$ is computable, but the series $\sum \alpha_n^2$ is not effectively convergent, and γ is a noncomputable real adjusted so that $\|f\| = 1$. Then:

$$\|x_n\|^2 = |\beta_n|^2 + \|y_n\|^2 + 2 \cdot \text{Re}\, [\bar\beta_n(f, y_n)].$$

We know that $\{\beta_n\}$, $\{|\beta_n|^2\}$, and $\{\|y_n\|^2\}$ are computable. Thus it suffices to show that

$$\{(f, y_n)\}$$

is computable. Let g be the projection of f on H_0, namely:

$$g = \sum_{n=1}^{\infty} \alpha_n e_n.$$

Then, since $y_n \in H_0$, $(f, y_n) = (g, y_n)$ for all n. Now g is not computable in H_0. However, the "Fourier coefficients" $\{\alpha_n\}$ of g are computable, and the sequence of vectors $\{y_n\}$ is computable. Hence, by Lemma 2, the sequence of inner products $\{(g, y_n)\}$ is computable. □

Lemma 4. *In terms of this ad hoc computability structure, T is effectively determined.*

Proof. Since T is bounded, it suffices to show that T acts effectively on the generating set $\{f, e_1, e_2, \ldots\}$ for H. We see at once that T acts effectively on the subspace H_0 generated by $e_1, e_2, \ldots$. Thus all we need to show is that Tf is computable. Now

$$Tf = \sum_{n=1}^{\infty} 2^{-n} \alpha_n e_n.$$

6. The Eigenvector Theorem, Completed

Since $Tf \in H_0$, we can apply Lemma 1. The sequence of "Fourier coefficients" $\{2^{-n}\alpha_n\}$ is computable, and the sum $\sum 4^{-n}\alpha_n^2$ is effectively convergent (being dominated by $\sum 4^{-n}$). Hence Tf is computable. □

We recall that the eigenvector of T corresponding to the eigenvalue 0 is Λ. Then we have:

Lemma 5. *The only constant multiple $c\Lambda$ of Λ which is ad hoc computable in H is the zero vector.*

Proof. We first note that if $c\Lambda$ is computable in H, then $|c|$ is a computable real. This follows from the Norm Axiom: for $\|c\Lambda\| = |c|$ must be computable if $c\Lambda$ is.

Now suppose $c\Lambda = \beta f + y$ is computable. Then the constant β and the vector y are computable. Furthermore, $f = \gamma\Lambda + \sum \alpha_n e_n$, where γ is the noncomputable real defined above. Since $y \in H_0$, the Λ-component of $\beta f + y$ is $\beta\gamma \cdot \Lambda$. Hence $c = \beta\gamma$, $|c| = |\beta|\gamma$ and $|c|$ is computable if and only if $\beta = 0$. □

As noted above, the preliminary Eigenvector Theorem is an immediate consequence of Lemmas 3, 4, and 5.

6. The Eigenvector Theorem, Completed

This section treats two main topics. One, as stated in its heading, is the completion of the Eigenvector Theorem. The second—and related—topic is ad hoc computability, as introduced in Chapter 2, Section 7. We prove a number of general results about ad hoc computability. These results are of independent interest, since they can be used as tools in a variety of situations. One of these situations is the proof of the Eigenvector Theorem.

The ad hoc computability results (Lemmas 7 and 8) are given in their most general form. This entails a bit of extra work. In particular, it requires the inclusion of Lemma 6, whose slightly complicated proof is deferred until Section 7. We mention that, for the proof of the Eigenvector Theorem, Lemma 6 could be omitted. But the most general forms of Lemmas 7 and 8, which apply to an *arbitrary* computability structure on a Hilbert space, require Lemma 6.

Although most of the results is this section apply only to Hilbert space, the following holds for Banach spaces.

Lemma 6 (The Effective Independence Lemma). *Let X be an effectively separable Banach space with an effective generating set $\{e_n\}$. Then there exists an effective generating set $\{f_n\}$ for X whose elements are linearly independent over the real or complex numbers.*

As mentioned above, the proof of this lemma is deferred until Section 7.

Lemma 7. *Let H be a Hilbert space with an arbitrary computability structure imposed upon it. Let $\{e_n\}$ be an effective generating set for H. Then there exists a computable orthonormal basis $\{u_n\}$ for H.*

Proof. By Lemma 6, we can assume without loss of generality that the elements of $\{e_n\}$ are linearly independent. The rest is easy. Following the standard Gram-Schmidt process, we write:

$$v_0 = e_0,$$

$$v_n = e_n - \sum_{k=0}^{n-1} \frac{(e_n, v_k)}{(v_k, v_k)} v_k.$$

Then the vectors v_n are orthogonal, and by the Linear Forms Axiom the sequence $\{v_n\}$ is computable.

[Strictly speaking, the inductive definition of $\{v_n\}$ given above does not fit the format of the Linear Forms Axiom. Namely, this axiom requires that we have linear forms in the original computable sequence $\{e_n\}$. However, we can compute the v_n in terms of the e_n by means of back-substitution.]

Finally, since the sequence of norms $\{\|v_n\|\}$ is computable by the Norm Axiom (and since $v_n \neq 0$ for all n, by linear independence), we can write

$$u_n = v_n/\|v_n\|,$$

giving a computable orthonormal sequence $\{u_n\}$. □

Remarks. Let us illustrate Lemma 7 as it applies to the Eigenvector Theorem. Begin with the effective generating set $\{f, e_1, e_2, \ldots\}$, which gave the ad hoc computability structure defined in Section 5. We recall that $\{e_1, e_2, \ldots\}$ is the standard computable orthonormal basis for the subspace H_0, but that f is an "exotic" element which is not computable in the standard sense. Now the Gram-Schmidt process begins with f, giving $u_0 = f$ (since $\|f\| = 1$). But f is *not* orthogonal to $e_1, e_2, \ldots$, and so the Gram-Schmidt process alters $e_1, e_2, \ldots$, producing a new orthonormal sequence $\{u_0, u_1, u_2, \ldots\}$. The sequence $\{u_n\}$ is not computable in the standard structure (for, as we have already seen, $u_0 = f$ is not computable in this sense). However, Lemma 7 implies that $\{u_n\}$ is computable in the ad hoc structure. Indeed, this is precisely the point of Lemma 7.

Now Lemma 8 below shows us how to translate results about the ad hoc computability structure into corresponding results for the standard structure. Although its proof is easy, it is a very useful result. In particular, it provides the final step in the proof of the Eigenvector Theorem.

Before turning to the technical details, we begin with a brief discussion of the content of Lemma 8. This Lemma asserts that, once we fix an orthonormal basis $\{u_k\}$ which is delared to be computable, the entire computability structure is fixed via conditions (i) and (ii) below. However, different computability structures can arise via the choice of different orthonormal bases. A basis which is computable in

6. The Eigenvector Theorem, Completed

terms of one computability structure might not be computable in terms of another. An example is the orthonormal basis $\{u_k\}$ above, which is ad hoc computable but not computable in the standard sense. Indeed, if the basis $\{u_k\}$ were computable in the standard sense, then by the Effective Density Lemma (Chapter 2, Section 5) the ad hoc computability structure would be the same as the standard one—which it clearly is not.

One final comment: Lemma 8 gives a representation theorem for any effectively separable computability structure on Hilbert space, and thus shows that all such structures are isomorphic. As we shall show in the next section, the corresponding statement for Banach spaces is false.

Lemma 8. *Consider any effectively separable computability structure—in the axiomatic sense—on a Hilbert space H. Then this structure has the following form. There is a computable orthonormal basis $\{u_k\}$ for H, and in terms of this basis: A sequence of vectors $\{x_n\}$ in H is computable if and only if*

i) *the double sequence of "Fourier coefficients" $\{c_{nk}\}$ of the x_n in terms of $\{u_k\}$ is computable, and*

ii) *the series $\sum_{k=0}^{\infty} |c_{nk}|^2$ converges effectively in k and n.*

Proof. Combine Lemma 1 from Section 5 with Lemma 7 above. □

Proof of the Eigenvector Theorem, completed. We combine the preliminary theorem of Section 5 with Lemma 8. Let T be the operator on the ad hoc space H given in Section 5. Let $u_0, u_1, u_2, \ldots$ be the ad hoc computable orthonormal basis for H (so that the ad hoc computability structure is given via Lemma 8). Let $e_0, e_1, e_2, \ldots$ be the standard computable orthonormal basis for $L^2[0, 1]$.

Let $\tilde{T}$ be the operator on $L^2[0, 1]$ which acts on the basis $\{e_0, e_1, e_2, \ldots\}$ in the same way that T acts on $\{u_0, u_1, u_2, \ldots\}$. That is, if

$$Tu = \sum_{k=0}^{\infty} b_{nk} u_k,$$

then

$$\tilde{T}e_n = \sum_{k=0}^{\infty} b_{nk} e_k.$$

What must we prove about $\tilde{T}$? Here let us recall the essential properties of T, as proved in Section 5. We showed there that T is compact and self-adjoint, and that 0 is an eigenvalue of T of multiplicity one. We showed further that T is effectively determined in terms of the ad hoc computability structure on H. Finally we showed that, in terms of this ad hoc structure, none of the eigenvectors of T corresponding to $\lambda = 0$ is computable.

We claim that $\tilde{T}$, acting on the standard computability structure, shares all of the properties of T listed above. These properties are of two types:

142   4. The Second Main Theorem, the Eigenvector Theorem, and Related Results

1) geometric properties—compactness, self-adjointness, the values and multiplicities of the eigenvalues, and
2) computability properties.

For (1), since $\{u_k\}$ and $\{e_k\}$ are both orthonormal bases, the transition from $\{u_k\}$ to $\{e_k\}$ preserves all of the above mentioned geometric properties. For (2), Lemma 8 quarantees that the transition from $\{u_k\}$ to $\{e_k\}$ maps the ad hoc computability structure on H onto the natural computability structure on $L^2[0, 1]$.

Thus $\tilde{T}$, acting on the standard computability structure, is the operator promised in the main Eigenvector Theorem. The proof of that theorem is now complete. □

One final question. What does the operator $\tilde{T}$ actually look like? Well, $\tilde{T}$ is just the operator T expressed in terms of the ad hoc computable orthonormal basis $\{u_k\}$. The basis $\{u_k\}$ is computed from the sequence $\{f, e_1, e_2, \ldots\}$ via Lemma 7, and the formula for f, on which this computation depends, is given in Section 5. Beginning with these observations, an explicit presentation of $\tilde{T}$ can be worked out. However, its form is rather a mess, and its intuitive meaning is completely hidden.

7. Some Results for Banach Spaces

For the first time in this chapter, we deal with computability structures on an arbitrary Banach space, rather than a Hilbert space. Our first result is the Effective Independence Lemma (Lemma 6 in Section 6), whose proof was postponed until this section. Our second result is a counterexample which shows that, on the Banach space l^1, there exist effectively separable ad hoc computability structures which are not isometric to the standard structure. This contrasts with the situation for Hilbert space (Lemma 8 in Section 6).

Effective Independence Lemma. *Let X be an effectively separable Banach space with an effective generating set $\{e_n\}$. Then there exists an effective generating set $\{f_n\}$ for X whose elements are linearly independent over the real or complex numbers.*

Proof of lemma. We begin with the effective generating set $\{e_n\}$, which we already have. The sequence $\{f_n\}$ will consist of a subset of $\{e_n\}$, selected by an effective process to be linearly independent and have dense linear span in X. We emphasize that $\{f_n\}$ is *not* a subsequence of $\{e_n\}$, since the terms in $\{f_n\}$ may appear in a different order than they do in $\{e_n\}$.

Proof sketch. The idea behind the construction of $\{f_n\}$ is roughly as follows. The construction proceeds by induction. At the q-th stage, suppose we already have the first $(k + 1)$ elements $f_0, \ldots, f_k$ of the desired sequence $\{f_n\}$. Then we examine in turn all of the elements $e_0, \ldots, e_q$ which do not already belong to the set $\{f_0, \ldots, f_k\}$, beginning with the e_i of smallest index and working upwards. For each e_i, we apply an effective "Test" (to be described below) which terminates in a finite number of

7. Some Results for Banach Spaces

steps and leads to one of the two conclusions: (a) the set $\{f_0, f_1, \ldots, f_k, e_i\}$ is linearly independent; (b) the vector e_i can be approximated to within a distance $<2^{-q}$ by a linear combination of $f_0, \ldots, f_k$. As soon as some e_i satisfies condition (a), we add e_i to $\{f_0, \ldots, f_k\}$; that is, we set $f_{k+1} = e_i$. In case several e_i satisfy (a), we select only one: that with the smallest i. When no e_i satisfies (a), we do nothing. Then we go on to the $(q + 1)$st stage.

By definition, the sequence $\{f_n\}$ so constructed is linearly independent. And by condition (b), any e_i which is forever omitted from $\{f_n\}$ can be arbitrarily closely approximated by linear combinations of the f_n; hence e_i lies in the closed linear span of $\{f_n\}$. Thus, since by definition the linear span of $\{e_n\}$ is dense in X, so is the linear span of $\{f_n\}$. This completes the proof-sketch.

Details of the proof. We begin by giving a criterion for the linear independence of any string $\{z_1, \ldots, z_k\}$ of computable vectors in X. This criterion will correspond to part (a) of the effective "Test" mentioned above.

By definition, the vectors $\{z_1, \ldots, z_k\}$ are linearly dependent if and only if there exist (real/complex) scalars $\{\beta_1, \ldots, \beta_k\}$, not all zero, such that $\beta_1 z_1 + \cdots + \beta_k z_k = 0$. Multiplying through by a constant, we can assume that $\{\beta_1, \ldots, \beta_k\}$ lies in the domain D between the unit sphere and the sphere of radius 2 in $\mathbb{R}^k$ or $\mathbb{C}^k$. Now we give a simple recipe for approximating all of the points in D by (dyadic rationals/dyadic complex rationals).

For $m = 0, 1, 2, \ldots$, let S_{mk} denote the set of all k-tuples $\{\beta_1, \ldots, \beta_k\}$ of (rationals/complex rationals) whose denominators are 2^m and which satisfy

$$1 \leq |\beta_1|^2 + \cdots + |\beta_k|^2 \leq 4.$$

Then, for each m and k, S_{mk} is a finite set, and there is an obvious procedure for listing S_{mk} for all m and k, effectively in m and k.

Independence Criterion. *The vectors $\{z_1, \ldots, z_k\}$ are linearly independent if and only if the following condition holds. For some $m \geq 2k$:*

$$\min \{\|\beta_1 z_1 + \cdots + \beta_k z_k\| : \{\beta_1, \ldots, \beta_k\} \in S_{mk}\} > 2^{-m} \cdot (\|z_1\| + \cdots + \|z_k\|).$$

Note. We observe that this criterion involves the evaluation of $\|\beta_1 z_1 + \cdots + \beta_k z_k\|$ at only a finite number of points, namely the points in the set S_{mk}. By the Linear Forms and Norm Axioms, $\|\beta_1 z_1 + \cdots + \beta_k z_k\|$ is computable, effectively in k, the β_i, and z_i. Similarly for $(\|z_1\| + \cdots + \|z_k\|)$.

Proof of criterion. For the "if" part. Suppose that the criterion holds, but that $\{z_1, \ldots, z_k\}$ are linearly dependent. We must derive a contradiction. Since $\{z_1, \ldots, z_k\}$ are dependent, there exists a point $\{\gamma_1, \ldots, \gamma_k\}$ on the sphere of radius $3/2$ in $\mathbb{R}^k$ or $\mathbb{C}^k$ (i.e. with $|\gamma_1|^2 + \cdots + |\gamma_k|^2 = 9/4$) such that $\gamma_1 z_1 + \cdots + \gamma_k z_k = 0$. Now for each γ_i, there exists a dyadic value β_i, with denominator 2^m, such that

$$|\gamma_i - \beta_i| \leq 2^{-m}.$$

[Actually, in the real case we can get $|\gamma_i - \beta_i| \leq (1/2)2^{-m}$; and in the complex case we can get a distance of $(\sqrt{2}/2)2^{-m}$.] Thus, since $\gamma_1 z_1 + \cdots + \gamma_k z_k = 0$,

$$\|\beta_1 z_1 + \cdots + \beta_k z_k\| \leq \sum_{i=1}^{k} |\gamma_i - \beta_i| \cdot \|z_i\| \leq 2^{-m}(\|z_1\| + \cdots + \|z_k\|).$$

This contradicts the inequality given in the criterion.

For the "only if" parts. If $\{z_1, \ldots, z_k\}$ are linearly independent, then over the entire domain $D = \{1 \leq |\beta_1|^2 + \cdots + |\beta_k|^2 \leq 4\}$ (without any reference to dyadic points), the minimum value δ,

$$\delta = \min \{\|\beta_1 z_1 + \cdots + \beta_k z_k\| : \{\beta_1, \ldots, \beta_k\} \in D\}$$

satisfies $\delta > 0$.

Now simply take any m large enough so that $2^{-m}(\|z_1\| + \cdots + \|z_k\|) < \delta$, and we see that the criterion does hold, as desired. This proves the validity of the criterion. □

Construction of $\{f_n\}$. Now we begin the construction of the sequence $\{f_n\}$. At this point we must take pains to work within the axioms for computability on a Banach space.

We start with the given effective generating set $\{e_n\}$. Then we sweep out the set of all finite (rational/complex rational) linear combinations of the e_n; this can be done in an effective way by using one of the standard recursive enumerations of all finite sequences of integers. By the Linear Forms Axiom, this yields a computable sequence $\{p_n\}$ in X which consists of all finite (rational/complex rational) linear combinations of the e_n. Then by the Norm Axiom, the sequence of norms $\{\|p_n\|\}$ is a computable sequence of reals.

We return now to the inductive definition of the sequence $\{f_n\}$, as described briefly in the proof-sketch above. We recall that at stage q we had selected $(k+1)$ linearly independent vectors $\{f_0, \ldots, f_k\}$ for $\{f_n\}$, and that we would then apply a "Test" to all of the elements $e_0, \ldots, e_q$ not already in the set $\{f_0, \ldots, f_k\}$. Here is the Test:

The Test. The test has two parts, (a) and (b), between which we alternate, switching back and forth until a termination is reached. The test is applied to an element e_i from the set $e_0, \ldots, e_q$.

Both of the parts (a) and (b) below involve strict inequalities (">" and "<" respectively). Hence, IF either of these inequalities holds, a finite amount of calculation (involving a sufficiently good rational approximation) will suffice to confirm it. The fact that this procedure halts will be proved below. As we shall see, (a) and (b) are not mutually exclusive. Indeed, this is why an effective decision procedure is possible.

Part (a). With reference to the "Independence Criterion" above: For $m = 2k, 2k+1, 2k+2, \ldots$, compute all of the values $\|\beta_0 f_0 + \cdots + \beta_k f_k + \beta_{k+1} e_i\|$ for the finite set of points $\{\beta_0, \ldots, \beta_{k+1}\} \in S_{m,k+2}$. Using rational approximations as discussed

7. Some Results for Banach Spaces

above, test whether:

$$\min \{\|\beta_0 f_0 + \cdots + \beta_k f_k + \beta_{k+1} e_i\|: \{\beta_0, \ldots, \beta_{k+1}\} \in S_{m,k+2}\}$$
$$> 2^{-m}(\|f_0\| + \cdots + \|f_k\| + \|e_i\|).$$

If this ever happens, for any m, cry "Halt!" and declare that "(a) holds: $\{f_0, \ldots, f_k, e_i\}$ are linearly independent."

Part (b). (This part is disappointingly unsubtle.) Scan through all of the (rational/complex rational) linear combinations of $f_0, \ldots, f_k$ until such a combination $\beta_0 f_0 + \cdots + \beta_k f_k$ is found for which

$$\|e_i - (\beta_0 f_0 + \cdots + \beta_k f_k)\| < 2^{-q}.$$

If this happens, cry "Halt!" and declare "(b) holds: e_i can be approximated to within a distance of 2^{-q} by a linear combination of $f_0, \ldots, f_k$."

Proof that "The Test" halts. By hypothesis, the vectors $\{f_0, \ldots, f_k\}$ are linearly independent. Therefore, either (A) the vectors $\{f_0, \ldots, f_k, e_i\}$ are independent, or (B') e_i is a linear combination of $f_0, \ldots, f_k$. By the "Independence Criterion", (A) is equivalent to (a) above. Now (B') is stronger than (b), but it suffices that (B') implies (b). Thus either (a) or (b) (or both) holds.

We still have to verify that one of the processes (a) or (b) halts. But this is easy. Both processes involve strict inequalities, ">" and "<" respectively. Thus IF either of the inequalities (a) or (b) is true, a finite amount of calculation will verify it, and the process will eventually halt. Since at least one of (a) and (b) is true, the process does halt, as desired. □

Now the rest of the proof is nearly identical to that given in the proof-sketch above. At stage q, we apply "The Test" to $e_0, \ldots, e_q$ (omitting those e_i already in $\{f_0, \ldots, f_k\}$). When the first e_i satisfies part (a) of the test, we add e_i to $\{f_0, \ldots, f_k\}$, setting $f_{k+1} = e_i$. Then we stop stage q. If no e_i satisfies part (a), we do nothing in stage q. Then we go on to stage $(q+1)$.

As explained in the proof-sketch above, the sequence $\{f_n\}$ is linearly independent and its closed linear span contains $\{e_n\}$; hence $\{f_n\}$ is dense in X.

Finally we must show that $\{f_n\}$ is computable in X. Now the construction of $\{f_n\}$ involved a recursive process, depending on the computable sequence of real numbers $\{\|p_n\|\}$. Thus f_n has the form $f_n = e_{a(n)}$ for a recursive function $a: \mathbb{N} \to \mathbb{N}$. By the Composition Property, proved as a consequence of the Axioms in Chapter 2, $\{f_n\}$ is computable. □

A counterexample for Banach spaces

Now that we have illustrated the usefulness of ad hoc computability structures, it seems natural to pose the following question. Are all such structures related via (not necessarily computable) isometries? We recall the definition:

Definition. Let X be a Banach space. A linear transformation $U: X \to X$ is called an *isometry* if U is onto and distance preserving (i.e. U is onto and $\|Ux\| = \|x\|$).

Of course, if U is an isometry, then U^{-1} exists and is an isometry.

At this point, we dispose of a triviality. Certain non-isometric mappings can give the same computability structure as an isometric mapping. For example, if U is an isometry and $k > 0$ is a computable real, then kU gives the same structure as U. Obviously such examples add nothing new to our problem, which as precisely formulated is:

Question. Are all effectively separable ad hoc computability structures $\mathscr{S}_1$ related to the standard structure $\mathscr{S}_0$ in the following way? There exists a (not necessarily computable) isometry U such that

$$\{f_n\} \in \mathscr{S}_1 \quad \text{if and only if } f_n = Ug_n \text{ for some } \{g_n\} \in \mathscr{S}_0.$$

All of the ad hoc computability structures which we have used so far have been based on non-computable isometries. Thus the two examples given in Chapter 2, Section 7 involved (1) multiplication by a non-computable complex constant c with $|c| = 1$, and (2) the translation $f(x) \to f(x + \alpha)$, where α is a noncomputable real. Both of these transformations are isometries. The more recondite example in Sections 5 and 6 involved an ad hoc computability structure on a Hilbert space. We proved in Lemma 8 above that any ad hoc computability structure on an effectively separable Hilbert space is isometric to the natural structure.

Suppose, then, that we turn our attention to Banach spaces. Here the situation is different.

Example. There exist ad hoc structures on l^1 which are not isometric to the standard structure.

To give such an example is not trivial. [The difficulty, of course, lies in satisfying the Norm Axiom. With the isometries—which preserved norms—this was no problem.] Since this example is used nowhere else in the book, we shall present it in a rather terse fashion. We remark that some of the steps are similar to those used in the Preliminary Eigenvector Theorem of Section 5.

Proof. We consider the real Banach space l^1, and let $e_0, e_1, \ldots$ deote the unit vectors, $e_n = \{0, 0, \ldots, 0, 1, 0, \ldots\}$ with a "1" in the n-th place. For typographical clarity we denote e_0 by Λ.

As usual, let $a: \mathbb{N} \to \mathbb{N}$ be a one to one recursive function generating a recursively enumerable nonrecursive set A. We assume that $0 \notin A$. Let

$$\alpha_n = 2^{-a(n)},$$

and set

$$\gamma = 1 - \sum_{n=1}^{\infty} \alpha_n.$$

7. Some Results for Banach Spaces

Then γ is a noncomputable real. Let

$$f = \gamma \Lambda + \sum_{n=1}^{\infty} \alpha_n e_n.$$

Then, by definition of γ, $\|f\| = 1$.

Let l_0^1 denote the subspace of l^1 spanned by $\{e_1, e_2, \ldots\}$.
We define the ad hoc computability structure on l^1 by specifying that

$$\{f, e_1, e_2, e_3, \ldots\}$$

is an effective generating set.

Equivalently, a sequence $\{x_n\}$ of vectors in l^1 is *ad hoc computable* if:

$$x_n = \beta_n f + y_n$$

where $\{\beta_n\}$ is a computable sequence of real numbers, and $\{y_n\}$ is a computable sequence of vectors (in the standard sense) in l_0^1.

Lemma. *This is a computability structure.*

Proof. The Linear Forms and Limit Axioms are clear. Only the Norm Axiom requires proof. That is, we must show that $\{\|x_n\|\}$ is computable if $\{x_n\}$ is. Fix n, and consider a single computable vector x. It will be clear that the procedure which follows is effective in n.

We have:

$$x = \beta f + y,$$

where β is a computable real number and y is computable in l_0^1.

Let $y = \{\theta_1, \theta_2, \ldots\}$, $\theta_i \in \mathbb{R}$.

Since y is computable in l^1, there exists a recursive function $e(N)$ such that

$$\sum_{k=e(N)}^{\infty} |\theta_k| \leq 2^{-N} \quad \text{for all } N.$$

To compute $\|x\|$ to within 2^{-N}, we use the following recipe:

Compute $\alpha_0, \alpha_1, \ldots, \alpha_{e(N)}$
Compute $\beta\alpha_0 + \theta_0, \ldots, \beta\alpha_{e(N)} + \theta_{e(N)}$.
Now

$$\|x\| = \sum_{k=1}^{\infty} |\beta\alpha_k + \theta_k| + |\beta|\gamma,$$

and to within an error of 2^{-N} (gotten by dropping the "tail" of $\{\theta_k\}$):

$$\|x\| = \sum_{k=1}^{e(N)} |\beta\alpha_k + \theta_k| + \sum_{k=e(N)+1}^{\infty} |\beta|\alpha_k + |\beta| \cdot \gamma.$$

The last two terms in the above displayed formula are not computable. However we claim that their sum

$$|\beta|\left(\gamma + \sum_{k=e(N)+1}^{\infty} \alpha_k\right)$$

is computable.

To show this, we reason as follows. Firstly, $|\beta|$ is computable. We know that

$$\gamma + \sum_{k=1}^{\infty} \alpha_k = 1,$$

so

$$\gamma + \sum_{k=e(N)+1}^{\infty} \alpha_k = 1 - \sum_{k=1}^{e(N)} \alpha_k. \quad \square$$

Lemma. *This structure is not isometric to the standard one.*

Proof. We recall the definition of an "extremal point" on the closed unit ball B of a Banach space. (Recall that $B = \{x: \|x\| \leq 1\}$.) We say that a vector $u \in B$ is *extremal* if there do not exist distinct vectors $v, w \in B$ and a constant c, $0 < c < 1$, such that $u = cv + (1-c)w$. Clearly any isometry preserves these extremal points.

It is well-known and easy to show that the only extremal points in l^1 are $\pm e_k$.

Now the standard computability structure on l^1 has an effective generating set consisting entirely of extremal points (namely the e_k, including $\Lambda = e_0$).

Thus, since isometries preserve extremal points, if the ad hoc structure were isometric to the standard one, the ad hoc structure would also have an effective generating set consisting entirely to extremal points. We now show that this cannot happen.

Suppose otherwise. The ad hoc effective generating set must contain some element z with a nonzero Λ-component. As we have seen, if z is extremal, then $z = \pm\Lambda$. On the other hand z—since it comes from an effective generating set—must be ad hoc computable. This is impossible in view of the following:

Lemma. *The only multiple of Λ which is ad hoc computable is $0 \cdot \Lambda$.*

Proof. Virtually identical to the proof of Lemma 5 in Section 5. $\quad \square$

This completes the proof of the previous lemma, and hence also of the main result. $\quad \square$

Chapter 5
Proof of the Second Main Theorem

Introduction

Here, as promised, we give the proof of the Second Main Theorem. (Cf. Chapter 4. The theorem is also restated at the end of this introduction.) For purposes of discussion, we recall two of the consequences of that theorem: The eigenvalues of an effectively determined self adjoint operator are computable, but the sequence of eigenvalues need not be.

How do we prove this? As might be expected, the proof is based on the spectral theorem. However, it does *not* involve an effectivization of that theorem. Nor does it involve an effectivization of some weaker version of that theorem. Rather we use certain consequences of the spectral theorem to develop an effective algorithm. This algorithm, in fact, embodies a viewpoint directly opposed to that of the spectral theorem—at least in its most standard form.

The standard form of the spectral theorem gives a decomposition of the Hilbert space H into mutually orthogonal subspaces $H_{(a_{i-1}, a_i]}$ corresponding to an arbitrary partition of the real line into intervals $(a_{i-1}, a_i]$. On these subspaces $H_{(a_{i-1}, a_i]}$ the operator T is "approximately well behaved". More precisely, (i) these subspaces are invariant under T—i.e. if x lies in the subspace, so does Tx, and (ii) the vectors x in the subspace are "approximate eigenvectors"—i.e. if $(a_{i-1}, a_i] \subseteq [\lambda - \varepsilon, \lambda + \varepsilon]$ then $\|Tx - \lambda x\| \leqslant \varepsilon \|x\|$.

It turns out that effective computations involving the spectral measure require the uniform norm, i.e. computability in the sense of Chapter 0. Thus the above decomposition—involving disjoint intervals—cannot be made effective. What we have is a classical analytic fact, the existence of such a decomposition, from which we must attempt to derive effective consequences. To do this we alter the standard spectral-theoretic decomposition in two ways.

First, we replace the disjoint intervals by intervals which overlap, after the manner of ... $[-2, 0]$, $[-1, 1]$, $[0, 2]$, $[1, 3]$, ... Second, we replace the characteristic functions of these intervals by "triangle functions" supported on them (cf. Pre-step B in Section 2). The overlapping intervals are necessary to account for the fact that a computable real number cannot be known exactly, and the triangle functions, being continuous, allow effective computations to be made. It is the necessity of using overlapping intervals and continuous functions which is at variance with the standard viewpoint of the spectral theorem—a viewpoint which stresses disjoint interval decompositions and orthogonal subspaces.

Returning to our original assertion: that the individual eigenvalues are computable but the sequence of eigenvalues need not be. Our overlapping triangles allow us to compute the individual eigenvalues. Essentially this depends on the fact that the eigenvalues occur at points where the spectral measure is "especially dense". By contrast, the sequence of eigenvalues need not be computable: this hinges on the fact that there is no effective way to distinguish between an eigenvalue and a very thin band of continuous spectrum.

The remarks given above are, of necessity, extremely brief. The same points will occur, more fully developed, at appropriate places throughout the proof. Cf. especially Section 3 (Heuristics).

We turn now to an outline of this chapter. In Sections 1–5 we will prove the Second Main Theorem for bounded self-adjoint operators. More precisely, we will prove the positive parts (i) and (ii) of that theorem. The extensions to normal and unbounded self-adjoint operators are given in Sections 6 and 7. Finally, in Section 8, we give the counterexamples required for the negative parts (iii) and (iv).

Remarks. Of course, it is the *unbounded* self-adjoint operators which—because of their applications in quantum mechanics and elsewhere—are the most interesting. However, the bounded case has to be done first, and it is there that the main difficulties lie. As we shall see, the extension to unbounded self-adjoint operators is rather straightforward once the bounded case has been proved.

As noted in Chapter 4, the proof of this theorem is long and arduous. For that reason, we have included a section on Heuristics (Section 3). The reader is advised to skim Sections 1 and 2, which give preliminary facts and definitions, and turn to Section 3 as soon as possible. Sections 4 and 5 spell out in rigorous detail the ideas sketched in Section 3.

Now for convenience, we restate the theorem. First we recall:

An unbounded operator $T: H \to H$ is called *effectively determined* if there is a sequence $\{e_n\}$ in H such that the pairs $\{(e_n, Te_n)\}$ form an effective generating set for the graph of T. In the case where T is bounded, this definition can be simplified— a fact that will prove useful. A *bounded* operator T is *effectively determined* if there is an effective generating set $\{e_n\}$ for H such that $\{Te_n\}$ is computable.

Second Main Theorem. *Let $T: H \to H$ be an effectively determined (bounded or unbounded) self-adjoint operator. Then there exists a computable sequence of real numbers $\{\lambda_n\}$ and a recursively enumerable set A of natural numbers such that:*

i) *Each $\lambda_n \in \mathrm{spectrum}(T)$, and the spectrum of T coincides with the closure of $\{\lambda_n\}$.*
ii) *The set of eigenvalues of T coincides with the set $\{\lambda_n: n \in \mathbb{N} - A\}$. In particular, each eigenvalue of T is computable.*
iii) *Conversely, every set which is the closure of $\{\lambda_n\}$ as in (i) above occurs as the spectrum of an effectively determined self-adjoint operator.*
iv) *Likewise, every set $\{\lambda_n: n \in \mathbb{N} - A\}$ as in (ii) above occurs as the set of eigenvalues of some effectively determined self-adjoint operator T. If the set $\{\lambda_n\}$ is bounded, then T can be chosen to be bounded.*

1. Review of the Spectral Theorem

This section presents those facts about the spectral theorem which are needed in this chapter. Nothing in it is new, nor has it anything to do with computability. The reader is advised to skim this section and return to it when necessary.

The basic results from spectral theory, which can be found in virtually any text on functional analysis, will be stated without proof. Any results that are not absolutely standard we derive. As references we mention Riesz, Sz.-Nagy [1955], Halmos [1951], Loomis [1953].

Such a survey, if it is to serve its purpose, cannot be too terse. Consequently we shall give a detailed review of the spectral theorem for bounded self-adjoint operators. Then we shall make a quick tour through the corresponding theories for normal and unbounded self-adjoint operators.

Let $T: H \to H$ be a bounded self-adjoint operator. Corresponding to T there is a "spectral measure" E, which is a mapping from Borel sets I in the line to operators E_I defined on H. More precisely: for each Borel subset I of the real line, E_I is the orthogonal projection onto a closed subspace H_I of H.

These subspaces H_I give a decomposition of the Hilbert space H which is a natural extension of the "eigenvector decomposition" for compact operators. (Compact self-adjoint operators, of course, form a special case—for which, as is well known, the eigenvectors are plentiful. On the other hand, we recall that a bounded self-adjoint operator may have no eigenvectors whatsoever. That is why the more complicated "spectral measure decomposition" is necessary.) The properties of the H_I, which generalize the elementary eigenvector situation, are:

(i) The subspaces H_I are invariant under T, i.e. $x \in H_I$ implies $T(x) \in H_I$.

(ii) For disjoint sets I, J in R, the spaces H_I and H_J are orthogonal.

(iii) If $I \subseteq [\lambda - \varepsilon, \lambda + \varepsilon]$ (in words, if I lies in a "thin" interval), then the $x \in H_I$ are "approximate eigenvectors"—more precisely, $\|Tx - \lambda x\| \leq \varepsilon \|x\|$.

The properties of the mapping $I \to H_I$ which justify the name "measure" are:

A. $H_{I \cap J} = H_I \cap H_J$
B. $H_{I \cup J} = H_I + H_J$ (direct sum)
C. (Countable additivity). If $I_1 \supseteq I_2 \supseteq \cdots$, and $\bigcap I_n = \emptyset$, then $\bigcap H_{I_n} = \{0\}$.

We recall that, in addition, the spectral measure is entirely supported on the spectrum of T, i.e. if $I \supseteq \text{spectrum}(T)$, then $H_I = H$.

We have already mentioned above that our construction in this chapter will require "triangle functions". For this reason we need to recall the "operational calculus" associated with the spectral theorem. We recall that, corresponding to any bounded real-valued Borel function f on $\mathbb{R}$, there is a bounded self-adjoint operator $f(T)$ represented by the "spectral integral"

$$f(T) = \int_{-\infty}^{\infty} f(t)\, dE(t),$$

where $dE(t)$ denotes integration with respect to the spectral measure. The use of the letter t, $-\infty < t < \infty$, reminds us that the domain of the spectral measure is the set of Borel subsets of $\mathbb{R}$.

As a special case, if we take the function $f(t) = t$, we obtain $f(T) = T$. That is, we obtain a representation for the operator T itself. Explicitly:

$$T = \int_{-\infty}^{\infty} t \, dE(t).$$

[Of course, the integration is really over spectrum(T), a compact set which supports the spectral measure. The function $f(t) = t$ is bounded on this set. We have written the integral as $\int_{-\infty}^{\infty}$ for simplicity.]

The following considerations, which are important in their own right, also shed light on the process of spectral integration. More importantly for our purposes, these results will be needed in this chapter.

The measure $d\mu_{xy}$

Let x, y be two vectors in H. It can be proved that, corresponding to x and y, there is a bounded complex measure $d\mu_{xy}$ with the following property. For all bounded real-valued Borel functions f:

$$(f(T)(x), y) = \int_{-\infty}^{\infty} f(t) \, d\mu_{xy}(t).$$

Here again, the use of the variable t, $t \in \mathbb{R}$, expresses the fact that the measure $d\mu_{xy}$ lies on the real line. We emphasize that $d\mu_{xy}$ (unlike dE) is an ordinary scalar-valued measure. Furthermore:

D. For $x = y$, the corresponding measure $d\mu_{xy}$ is positive (Sometimes we write $d\mu(x)$ for $d\mu_{xx}$.)

The operational calculus

The operational calculus—i.e. the mapping from functions f to operators $f(T)$ described above—will play a key role in the proof of our Second Main Theorem. In fact, as an algorithmic tool, the operational calculus is very powerful. Its power resides in the fact that it gives a natural isomorphism between the "arithmetic" of functions and the "arithmetic" of operators.

In what follows, we assume that f and g are bounded real-valued Borel functions; α and β are real scalars; and I denotes a Borel subset of $\mathbb{R}$.

E. (Linearity). $(\alpha f + \beta g)(T) = \alpha f(T) + \beta g(T)$.
F. (Multiplication). $(fg)(T) = f(T) \cdot g(T)$.
G. (Boundedness). $\|f(T)\| \leq \sup \{|f(\lambda)|: \lambda \in \text{spectrum}(T)\}$.
H. (Pointwise convergence). Let $\{f_n\}$ be a uniformly bounded sequence of Borel functions such that, as $n \to \infty$, $f_n(t) \to 0$ for all t. Then, for any vector x in H,

$$f_n(T)(x) \to 0 \quad \text{in the norm of } H.$$

1. Review of the Spectral Theorem

I. (Projections). Let $\chi = \chi_I$ be the characteristic function of I. Then $\chi(T)$ coincides with the projection E_I given by the spectral measure.

This completes our list of standard results. We remind the reader that proofs can be found e.g. in Riesz, Sz.-Nagy [1955], Halmos [1951], Loomis [1953].

Technical Corollaries

We now reach a transitional stage in this introductory section. The above results are standard, but they are not in the form that we need. The following list contains precisely those consequences of the spectral theorem that we will use in proving the Second Main Theorem. There are seven of them.

Remarks. Of course, all of the results in this section are known to specialists. The results A–I above are absolutely standard, and we refer the reader to the literature for their proofs. However, some of the results below are harder to locate in the textbook literature. They are necessary for our proof, and so, for the convenience of the reader, we shall work them out.

To aid the reader in skimming through the derivations which follow, it is useful to stress that two results, taken from the list A–I above, will be used repeatedly.

1) The operational calculus is multiplicative: $(fg)(T) = f(T)g(T)$.
2) Characteristic functions correspond to projections: if $\chi = \chi_I$ is the characteristic function of a Borel set $I \subseteq \mathbb{R}$, then $\chi(T)$ is the associated projection E_I on the subspace H_I.

We turn now to the seven corollaries of the Spectral Theorem that we need for our proof. These fall into four categories, wich we have put under appropriate headings.

Criteria for nullity

SpThm 1. *Let I be a Borel set in $\mathbb{R}$, and let x be a vector in H_I. Let f be a bounded Borel function such that $\text{support}(f) \cap I = \emptyset$. Then $f(T)(x) = 0$.*

Proof. Let $\chi = \chi_I$ be the characteristic function of I. Then $\chi(T)$ is the projection on H_I. Since $x \in H_I$, $\chi(T)(x) = x$. Now f and χ have disjoint supports, so that $f\chi = 0$. Hence $0 = (f\chi)(T)(x) = f(T)\chi(T)(x) = f(T)(x)$. □

SpThm 2 (Pointwise convergence almost everywhere). *Let I be a Borel set for which H_I is the zero subspace. Let $\{f_n\}$ be a uniformly bounded sequence of Borel functions which is pointwise convergent to zero except on I: i.e. $f_n(t) \to 0$ as $n \to \infty$ for all $t \notin I$. Then, for any vector x, $f_n(T)(x) \to 0$.*

Proof. Let $\chi = \chi_I$ be the characteristic function of I, so that $\chi(T)$ is the projection on H_I. Since H_I is null, $\chi(T) = 0$. Hence $(1 - \chi)(T) =$ identity operator. Now for all real t:

$$f_n(t)(1 - \chi(t)) \to 0 \quad \text{as} \quad n \to \infty,$$

since $f_n(t) \to 0$ for $t \notin I$, and $1 - \chi(t) = 0$ for $t \in I$. Hence by H. above,

$$[f_n(1 - \chi)](T)(x) \to 0 \quad \text{as} \quad n \to \infty.$$

By the multiplicative property of the operational calculus, we deduce:

$$f_n(T)(1 - \chi)(T)(x) \to 0 \quad \text{as} \quad n \to \infty.$$

But since $(1 - \chi)(T) =$ identity, this means that $f_n(T)(x) \to 0$. □

The question of whether or not certain subspaces H_I are null will play an important role in the proof of the Second Main Theorem. For example, even if a point $\lambda \in \text{spectrum}(T)$, the corresponding subspace $H_{\{\lambda\}}$ may be null. However, this cannot happen for a neighborhood $(\lambda - \varepsilon, \lambda + \varepsilon)$ of λ. Furthermore, we will show that $H_{\{\lambda\}}$ itself is non-null if and only if λ is an eigenvalue of T. The key results here are SpThm 3 and SpThm 5 below.

SpThm 3. *Let $\lambda \in \text{spectrum}(T)$. Then, for any $\varepsilon > 0$, the subspace $H_{(\lambda - \varepsilon, \lambda + \varepsilon)}$ is nonzero.*

Proof. Suppose otherwise. Let χ be the characteristic function of $(\lambda - \varepsilon, \lambda + \varepsilon)$, so that $\chi(T)$ is the corresponding projection. By assumption, $\chi(T) = 0$, so that $(1 - \chi)(T) =$ identity operator. Let

$$g(t) = \begin{cases} (t - \lambda)^{-1} & \text{for } t \notin (\lambda - \varepsilon, \lambda + \varepsilon), \\ 0 & \text{otherwise.} \end{cases}$$

Since $(\lambda - \varepsilon, \lambda + \varepsilon)$ is a neighborhood of λ, the function g is bounded. Hence $g(T)$ is a bounded operator.

We now show that $g(T) = (T - \lambda)^{-1}$. Consider the corresponding functions of a real variable. We have:

$$(t - \lambda)g(t)(1 - \chi(t)) = 1 - \chi(t),$$

since $g(t) = (t - \lambda)^{-1}$ for $t \notin (\lambda - \varepsilon, \lambda + \varepsilon)$, and $1 - \chi(t) = 0$ for $t \in (\lambda - \varepsilon, \lambda + \varepsilon)$. Hence, again by the multiplicative property of the operational calculus,

$$(T - \lambda)g(T)(1 - \chi(T)) = 1 - \chi(T).$$

But we have seen that $1 - \chi(T) =$ identity operator. Thus $(T - \lambda)g(T) =$ identity. Similarly one shows that $g(T)(T - \lambda) =$ identity. Hence $g(T)$ is an inverse to $(T - \lambda)$, contradicting the fact that $\lambda \in \text{spectrum}(T)$. □

Eigenvalues

Recall that, by definition, λ is an "eigenvalue" of T if there is some "eigenvector" $x \neq 0$ with $Tx = \lambda x$. The eigenvalues form a subset of the spectrum.

1. Review of the Spectral Theorem

SpThm 4. *Let λ be an eigenvalue of T with eigenvector x. Then, for any continuous function f,*
$$f(T)(x) = f(\lambda) \cdot x.$$

[Actually, the same thing holds for bounded Borel functions f. This more difficult result is an easy consequence of SpThm 5 below, but we have no need of it.]

Proof. Since x is an eigenvector for λ, $Tx = \lambda x$, whence $T^n x = \lambda^n x$, whence $p(T)(x) = p(\lambda) \cdot x$ for any polynomial p. Now the result extends to continuous functions f by the Weierstrass Approximation Theorem, combined with G. above. □

SpThm 5. *A vector x is an eigenvector for λ if and only if $x \in H_{\{\lambda\}}$, where $H_{\{\lambda\}}$ is the subspace corresponding to the point-set $\{\lambda\}$. In particular, λ is an eigenvalue if and only if $H_{\{\lambda\}}$ is nonzero.*

Proof. The "if" part is trivial. If $x \in H_{\{\lambda\}}$, then in the spectral measure $dE(t)$, x belongs exclusively to the part where $t = \lambda$. Thus in the integral representation for Tx.

$$Tx = \int t \cdot dE(t)(x),$$

only the value $t = \lambda$ is relevant, and x is multiplied by λ.

The converse is a little harder. Let x be an eigenvector for λ. Without loss of generality, we can asume that $\lambda = 0$. We use the "triangle functions" τ_n (see Figure 0) defined by the equations:

$$\tau_n(t) = \begin{cases} 1 - nt & \text{for } 0 \leq t \leq 1/n, \\ 1 + nt & \text{for } -1/n \leq t \leq 0, \\ 0 & \text{for } |t| \geq 1/n. \end{cases}$$

Let δ be the characteristic function of $\{0\}$ (i.e. $\delta(t) = 1$ if $t = 0$, $\delta(t) = 0$ otherwise).

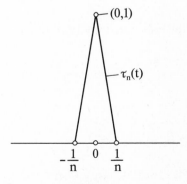

Figure 0

Then as $n \to \infty$, the functions $\tau_n(t)$ converge pointwise to $\delta(t)$. Hence by H. above:

$$\tau_n(T)(x) \to \delta(T)(x).$$

Since x is an eigenvector for $\lambda = 0$, and since the functions τ_n are continuous, $\tau_n(T)(x) = \tau_n(0) \cdot x = x$ by SpThm 4. Hence, since $\tau_n(T)(x) \to \delta(T)(x)$, $\delta(T)(x) = x$. But $\delta(T)$ is just the projection onto the subspace $H_{\{0\}}$. Hence $x \in H_{\{0\}}$. □

The measure $d\mu(x)$

Recall the complex measure $d\mu_{xy}$ discussed above. If we set $x = y$, then we obtain a *positive* measure $d\mu_{xx} = d\mu(x)$, determined by the vector x.

In integration formulas, we may want to display the real variable t: then we write $d\mu(x) = d\mu(x, t)$. As we saw above for $d\mu_{xy}$, the defining equation for $d\mu_{xx} = d\mu(x)$ is:

$$(f(T)(x), x) = \int_{-\infty}^{\infty} f(t) \, d\mu(x, t),$$

where f is any bounded Borel function, and $f(T)$ is the corresponding operator.

SpThm 6. *Let I be any Borel set in $\mathbb{R}$. Let x be any vector in H, and let x_0 be the projection of x on H_I. Then:*

$$\|x_0\|^2 = \text{the } d\mu(x)\text{-measure of } I.$$

(In particular, $\|x\|^2 = $ the $d\mu(x)$-measure of $\mathbb{R}$.)

Proof. Let $\chi = \chi_I$ be the characteristic function of I, so that $\chi(T)$ is the projection on the subspace H_I. Then:

$$\|x_0\|^2 = \|\chi(T)(x)\|^2 = (\chi(T)(x), \chi(T)(x))$$
$$= (\chi(T)^2(x), x) = (\chi^2(T)(x), x) = (\chi(T)(x), x),$$

since $\chi^2 = \chi$. Now by the defining equation for $d\mu(x)$, this becomes:

$$\int_{-\infty}^{\infty} \chi(t) \, d\mu(x, t) = \int_I d\mu(x, t) = \text{the } d\mu(x)\text{-measure of } I. \quad \square$$

Uniform approximation

SpThm 7. *For any bounded Borel function f,*

$$\|f(T)\| \leq \sup\{|f(\lambda)| : \lambda \in \text{spectrum}(T)\}.$$

1. Review of the Spectral Theorem

Proof. This is just the standard fact G. above. We have restated it here because we promised to list (in the SpThm N category) every spectral theoretic result needed for the proof of the Second Main Theorem. This result is the last on our list. □

Bounded Normal and Unbounded Self-adjoint Operators

Here, as promised above, we shall be brief. We recall that a bounded operator $T: H \to H$ is said to be *normal* if $TT^* = T^*T$. All of the above results extend to bounded normal operators, once the following trivial modifications are made:

a) Whereas the spectrum of a self-adjoint operator is real, the spectrum of a bounded normal operator is a compact subset of the complex plane. Hence, for our Borel sets I, we take Borel subsets of the complex plane.
b) Similarly, in the operational calculus, we consider complex-valued (as opposed to real-valued) functions. We continue to assume that these functions are bounded and Borel. Then all of the identities A–I above continue to hold, and there is one new entry on the list. If $\bar{f}$ denotes the complex conjugate of the function f, and "*" denotes adjoint, then:

J. $$\bar{f}(T) = [f(T)]^*.$$

It follows that if f is real-valued, then the operator $f(T)$ is self-adjoint. In particular, the projections $\chi(T)$ are self-adjoint (since the values of the characteristic function $\chi = \chi_I$ are real, whether or not the set I lies within the real line). Of course, the operator T itself need not be self-adjoint, because the function $f(z) = z$ (corresponding to $f(T) = T$) is not real-valued in the complex plane.

The proofs of A–J for bounded normal operators can be found in standard references (e.g. Riesz, Sz.-Nagy [1955], Halmos [1951]). The corresponding extensions of SpThm 1–SpThm 7 are then obvious: Again, complex Borel sets I and functions f replace the real sets/functions discussed above. In SpThm 3, a disk $\{z: |z - \lambda| < \varepsilon\}$ in the complex plane replaces the real interval $(\lambda - \varepsilon, \lambda + \varepsilon)$. Otherwise, the statements of SpThm 1–SpThm 7 for normal operators are identical to those given above for self-adjoint operators. The proofs are so similar to those already given that we leave them to the reader.

Unbounded self-adjoint operators

We recall from the introduction to Chapter 4 that the adjoint of an unbounded closed operator is defined via its graph, and an unbounded operator T is said to be *self-adjoint* if it coincides with its graph-theoretic adjoint.

Only one result concerning unbounded self-adjoint operators will be needed in this book. Most textbook presentations of operator theory give this result as a lemma (see e.g. Riesz, Sz.-Nagy [1955]). The result is:

Proposition. *Let T be a (bounded or unbounded) self-adjoint operator. Then the inverse $(T - i)^{-1}$ exists and is a bounded normal operator.*

2. Preliminaries

This section presents a number of technical definitions and results which are needed before we can come to the core of the Second Main Theorem. The reader who prefers a broad overview may wish to skim this section and then turn directly to Section 3 (Heuristics). The topics in this section are presented in the order in which they occur in the proofs. However, for skimming purposes, the most important subsection is Pre-step B (the triangle functions). We begin with:

Lemma (Uniformity in the exponent). *Let X be a Banach space with a computability structure, and suppose that X has an effective generating set $\{e_n\}$. Let $T: X \to X$ be an effectively determined bounded linear operator. (Since T is bounded, the hypothesis "effectively determined" means simply that $\{Te_n\}$ is computable.) Then the double sequence $\{T^N e_n\}$ is computable in both variables N and n.*

Proof. At first glance, this would appear to be a simple induction. The difficulty is to give a proof which stays within the axioms for computability on a Banach space (cf. Chapter 2, Section 1). This difficulty is resolved by extending the Effective Density Lemma of Chapter 2, Section 5, in a manner which reduces the problem to multilinear algebra. By hypothesis, $\{Te_n\}$ is computable. Hence, by the Effective Density Lemma, there is a computable triple sequence $\{\alpha_{nkj}\}$ of real/complex *rationals* and a recursive function $d(n, k)$ such that: If we write

$$p_{nk} = \sum_{j=0}^{d(n,k)} \alpha_{nkj} e_j,$$

then

$$\|p_{nk} - Te_n\| \leq 2^{-k} \qquad \text{for all } n, k.$$

We also observe that since T is bounded, there is an integer C such that $\|T\| \leq C$. Without loss of generality, we can replace T by $T/2C$, and thus assume that $\|T\| \leq 1/2$.

To prove the lemma, we shall construct a computable 4-fold sequence $\{\beta_{Nnkj}\}$ of (real/complex) *rationals* and a recursive function $e(N, n, k)$ such that: If we write

$$q_{Nnk} = \sum_{j=0}^{e(N,n,k)} \beta_{Nnkj} e_j,$$

then

$$\|q_{Nnk} - T^N e_n\| \leq 2^{-k} \qquad \text{for all } N, n, k.$$

This is done by an induction on the (real/complex) *rational* coefficients β_{Nnkj}, using the sequence $\{\alpha_{nkj}\}$ which we already have. The process operates strictly within the domain of integers and their quotients.

2. Preliminaries

We define β_{Nnkj} by induction on N.

For $N = 0$, we set $\beta_{0nkj} = 1$ if $j = n$, 0 otherwise. (This gives $q_{0nk} = e_n$ for all k.) Assume that β_{Nnkj} is defined for a fixed N, and all n, k, j. We now define $\beta_{N+1,nkj}$.

[Recall that $\|T\| < 1/2$. Thus from the inductive assumption that $\|q_{Nnk} - T^N e_n\| < 2^{-k}$, we deduce that $\|Tq_{Nnk} - T^{N+1} e_n\| < 2^{-k}/2$. Now we examine q_{Nnk} with a view towards approximating Tq_{Nnk}.]

Consider

$$q_{Nnk} = \sum_{j=0}^{e(N,n,k)} \beta_{Nnkj} e_j.$$

Each β_{Nnkj} is a real or complex rational; let D_{Nnkj} be the least integer greater than $|\beta_{Nnkj}|$. Let

$$E_{Nnk} = \sum_{j=0}^{e(N,n,k)} D_{Nnkj}.$$

Let $s = s(N, n, k)$ be the least integer such that $2^{-s} \leq 2^{-k}/2E_{Nnk}$.

Now we define $q_{N+1,nk}$ by substituting p_{js} for e_j in the formula for q_{Nnk}:

$$q_{N+1,nk} = \sum_{j=0}^{e(N,n,k)} \beta_{Nnkj} p_{js}.$$

[By the manner in which p_{js} approximates Te_j, we have $\|p_{js} - Te_j\| < 2^{-s}$. Hence by the definition of D_{Nnkj}, E_{Nnk}, and s, we have $\|q_{N+1,nk} - Tq_{Nnk}\| < 2^{-k}/2$.]

Now in the above sum, we replace the index j by i, and then put in the definition of p_{is}:

$$q_{N+1,nk} = \sum_{i=0}^{e(N,n,k)} \beta_{Nnki} \sum_{j=0}^{d(i,s)} \alpha_{isj} e_j.$$

Thus we define:

$$\beta_{N+1,nkj} = \sum_{i=0}^{e(N,n,k)} \beta_{Nnki} \cdot \alpha_{isj}, \qquad s = s(N, n, k).$$

The new limit of summation $e(N + 1, n, k)$ is the maximum j for which the above double sum is nonempty, i.e.

$$e(N + 1, n, k) = \max \{d(i, s): 0 \leq i \leq e(N, n, k)\}.$$

This completes the definition of the multi-sequence $\{\beta_{Nnkj}\}$.

Now we return to the Banach space X. We must show that the desired inequality,

$$\|q_{Nnk} - T^N e_n\| < 2^{-k},$$

extends by induction on N to all N, n, k. For $N = 0$ it is trivial. Assume that it holds for N. On this assumption, we have already seen (in the bracketed remarks above) that $\|Tq_{Nnk} - T^{N+1}e_n\| < 2^{-k}/2$, and $\|q_{N+1,nk} - Tq_{Nnk}\| < 2^{-k}/2$. Combining these two inequalities gives the desired result.

Now that we have constructed $\{\beta_{Nnkj}\}$ and $\{q_{Nnk}\}$ with the desired properties, the rest is easy. The Linear Forms Axiom implies that $\{q_{Nnk}\}$ is computable in X, and the Limit Axiom implies that $\{T^N e_n\}$ is computable. □

Corollary. *Let* $T: X \to X$ *be bounded and effectively determined, and let* $\{y_n\}$ *be a computable sequence in* X. *Then* $\{T^N y_n\}$ *is computable, effectively in* N *and* n.

Proof. Since $\{y_n\}$ is computable, the Effective Density Lemma asserts that there is a computable double sequence $r_{nk} = \sum \alpha_{nkj} e_j$ such that $\|r_{nk} - y_n\| \to 0$ as $k \to \infty$, effectively in k and n. Since T is bounded and effectively determined, the preceding lemma tells us that $\{T^N e_n\}$ is computable, effectively in N and n. Now the Linear Forms Axiom implies that $\{T^N r_{nk}\}$ is computable, effectively in all variables. Finally, since T is bounded, $\|T^N r_{nk} - T^N y_n\| \to 0$ as $k \to \infty$, effectively in all variables. Hence by the Limit Axiom, $\{T^N y_n\}$ is computable. □

The interval $[-M, M]$

Since T is a bounded self-adjoint operator, spectrum(T) is a compact subset of the real line. We take an integer M such that

$$\text{spectrum}(T) \subseteq [-(M-1), M-1],$$

and then work within the interval $[-M, M]$ in order to give ourselves "room around the edges". Throughout the remainder of this proof, M designates the fixed integer defined above.

The sequence $\{x_n\}$

We recall that H is an effectively separable Hilbert space with an effective generating set $\{e_n\}$. In this proof we will use a computable sequence of vectors $\{x_n\}$ such that:

$$1 < \|x_n\| < 1001/1000 \qquad \text{for all } n,$$

and

$$\{x_n\} \text{ is dense on the annulus } \{x: 1 \leqslant \|x\| \leqslant 1001/1000\}.$$

It is important to stress that, by these hypotheses, $\|x_n\| > 1$. Furthermore, the closure of $\{x_n\}$ in H contains the unit sphere $\{x: \|x\| = 1\}$.

The construction of $\{x_n\}$ is very simple. We begin by taking the sequence $\{x'_n\}$ of all (real/complex) rational linear combinations of the elements e_n in the effective generating set. Then $\{x'_n\}$ is computable by the Linear Forms Axiom. Next we effectively list (not necessarily in their original order) all of the elements of $\{x'_n\}$ which

2. Preliminaries

satisfy $1 < \|x'_n\| < 1001/1000$. For the sake of completeness, we indicate precisely how this is done.

By the Norm Axiom, since $\{x'_n\}$ is computable, the norms $\{\|x'_n\|\}$ form a computable sequence of real numbers. Hence there is a computable double sequence of rational approximations R_{nk} with $|R_{nk} - \|x'_n\|| \leq 1/2^k$ for all n, k. Now we effectively scan the double sequence $\{R_{nk}\}$, using a procedure which returns to each n infinitely often. Whenever an R_{nk} shows up with $1 + (1/2^k) < R_{nk} < (1001/1000) - (1/2^k)$, we add the corresponding vector x'_n to our list. In this way we eventually find all of those x'_n, and only those x'_n, which satisfy $1 < \|x'_n\| < 1001/1000$.

The resulting list is the desired computable sequence $\{x_n\}$.

The constructions of the interval $[-M, M]$ and the sequence $\{x_n\}$, while essential, were rather elementary. The three "Pre-steps" which follow are somewhat more elaborate.

Pre-step A (the effective operational calculus)

Here we must find the effective content of the operational calculus, as laid out (noneffectively) in Section 1. More precisely, we must develop—as corollaries of the spectral theorem—operations which can be made effective and which will allow us to proceed with our construction.

We begin with the assumption, made in the Second Main Theorem, that T is an effectively determined self-adjoint operator. Here and until the end of Section 5, we also assume that T is bounded. Then from the above lemma (Uniformity in the exponents) and its corollary, we have:

Let $\{y_n\}$ be a computable sequence of vectors in H. Then the double sequence $\{T^N y_n\}$ is computable in H, effectively in N and n.

Now, in terms of the operational calculus, T^N is just the action of the function $f(t) = t^N$ on the operator T; i.e. if $f(t) = t^N$, then $f(T) = T^N$.

[For completeness we give the proof. Since $T = \int_{-\infty}^{\infty} t \, dE(t)$, T itself corresponds to the function $f(t) = t$. Then the extension to powers of t (or T) follows from the multiplicative law: $(fg)(T) = f(T)g(T)$.]

Now by the Linear Forms Axiom (cf. Chapter 2) the above extends immediately to any computable sequence of polynomials. Thus we have:

Let $\{y_n\}$ be a computable sequence of vectors in H, and let $\{p_m\}$ be a computable sequence of polynomials. Then the double sequence $\{p_m(T)(y_n)\}$ is computable in H.

Lemma. *Let $[-M, M]$, $M =$ integer, be an interval containing spectrum(T). Let $\{f_m\}$ be a sequence of continuous functions on $[-M, M]$ which is computable in the sense of Chapter 0. Let $\{y_n\}$ be a computable sequence of vectors in H. Then $\{f_m(T)(y_n)\}$ is a computable double sequence of vectors in H.*

Proof. We use the result from spectral theory (SpThm 7) that, for any bounded Borel function f, the operator norm

$$\|f(T)\| \leq \sup\{|f(\lambda)|: \lambda \in \text{spectrum}(T)\}.$$

We also use the "Weierstrass approximation" variant of the notion of a computable sequence of continuous functions $\{f_m\}$. (Cf. Section 3 in Chapter 0). By this definition, there is a computable double sequence of polynomials $\{p_{mk}\}$ which converges uniformly to f_m as $k \to \infty$, effectively in k and m.

The rest is easy. A uniform bound on $|f_m(t) - p_{mk}(t)|$ gives (by SpThm 7) the same bound on the uniform operator norm $\|f_m(T) - p_{mk}(T)\|$. We already know that we can compute $\{p_{mk}(T)(y_n)\}$, effectively in m, k, and n. Now we apply the Limit Axiom (Chapter 2): the uniform convergence in operator norm implies the computability of $\{f_m(T)(y_n)\}$, as desired. □

Corollary. *With $\{f_m\}$ and $\{y_n\}$ as above, the sequence of norms*

$$\|f_m(T)(y_n)\|$$

is computable, effectively in both m and n.

Proof. This follows immediately from the above lemma, together with the Norm Axiom of Chapter 2. □

Notes. These arguments break down if we attempt to deal with $f(T)$ for discontinuous functions f. For then we would have to deal with pointwise rather than uniform convergence, a notion that is frequently not effective.

Pre-step A involves a triple transition from continuous functions f_m to operators $f_m(T)$ to vectors $f_m(T)(y_n)$ to norms $\|f_m(T)(y_n)\|$. This is quite natural, since to compute an operator means to compute its action on vectors, and the easiest thing to compute about a vector is its norm.

Pre-step B (the triangle functions)

As stated above, in order to obtain computability in our application of the spectral theorem, we must work with continuous functions. On the other hand, we want to preserve—so far as is possible—the idea of a decomposition of the interval $[-M, M]$ into subintervals. This is achieved by using triangle functions (definitions to follow). The supports of these triangle functions overlap, in the manner of

$$\ldots [-2, 0], [-1, 1], [0, 2], [1, 3], \ldots$$

Our construction will proceed in stages, indexed by $q = 0, 1, 2, \ldots$. At the 0-th stage, we pave the interval $[-M, M]$ with overlapping intervals of length 2, as displayed above. Then we subdivide these intervals, reducing the mesh by a factor of $1/8$ at each stage. At the q-th stage, we have overlapping intervals of length $2 \cdot 8^{-q}$, the i-th such interval being

$$I_{qi} = [(i-1)8^{-q}, (i+1)8^{-q}], \quad \text{where } -M \cdot 8^q < i < M \cdot 8^q.$$

2. Preliminaries

Now the corresponding triangle function τ_{qi}, whose support is I_{qi}, is given by:

$$\tau_{qi} = \tau(8^q x - i), \qquad -M \cdot 8^q < i < M \cdot 8^q,$$

where

$$\tau(x) = \begin{cases} 1 - |x| & \text{for } |x| \leq 1, \\ 0 & \text{elsewhere.} \end{cases}$$

We observe that the triangle functions τ_{qi} are symmetrical and rise to a peak at the midpoints of the intervals I_{qi}. (Cf. Figures 1 and 2.)

At the initial stage in our construction, which we call the -1-st stage, we do not use triangle functions. Instead we use a trapezoidal function σ such that $\sigma(x) = 1$ on $[-(M-1), M-1]$, and $\sigma(x)$ drops linearly to zero at $\pm M$. (See Figure 2.) We observe that, since spectrum$(T) \subseteq [-(M-1), M-1]$, $\sigma(x)$ is identically equal to 1 on the spectrum of T.

We shall need an identity which shows how each $\tau_{q-1,j}$ decomposes into triangle functions τ_{qi} of the next generation. Consider a fixed $\tau_{q-1,j}$. To conform with later notations, we shall denote this fixed $\tau_{q-1,j}$ by τ^*_{q-1}. Similarly the interval $I_{q-1,j}$ will be written I^*_{q-1}. Finally we set $h = 8j$.

Note. In the body of the proof, τ^*_{q-1} will be a particular one of the $\tau_{q-1,j}$, chosen via an inductive process. The identities of this subsection hold for *any j*, and hence they hold for the *particular j* which we eventually select.

As a preface to the first identity, we make some geometric observations. Contained within the interval τ^*_{q-1}, there are precisely fifteen subintervals of the q-th generation, namely

$$I_{q,h-7}, \ldots, I_{q,h+7} \qquad (h = 8j).$$

We shall decompose the triangle function τ^*_{q-1} into a linear combination of the triangle functions τ_{qi}, $h - 7 \leq i \leq h + 7$. This is done as follows:

$$\tau^*_{q-1}(x) = \tfrac{1}{8}[\tau_{q,h-7}(x) + 2 \cdot \tau_{q,h-6}(x) + \cdots + 7 \cdot \tau_{q,h-1}(x) + 8 \cdot \tau_{q,h}(x)$$
$$+ 7 \cdot \tau_{q,h+1}(x) + \cdots + 2 \cdot \tau_{q,h+6}(x) + \tau_{q,h+7}(x)]$$

(See Figure 1.)

Proof. For the sake of completeness, we prove the above identity. The easiest proof is via slopes. Firstly, all of the functions in the above identity are continuous. Therefore is suffices to show that both sides of the equation have the same slope *at all non-partition points* $x \neq i \cdot 8^{-q}$ (i.e. at all points which are not vertices of the triangles τ_{qi}).

Without loss of generality, we can consider the left hand side of the "big" triangle τ^*_{q-1}, i.e. the region where τ^*_{q-1} has positive slope. On this region, the slope of τ^*_{q-1}

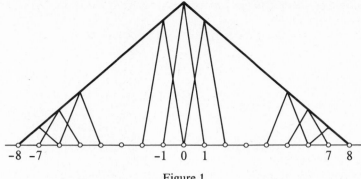

Figure 1

is 8^{q-1}. By contrast, the triangles τ_{qi} have slopes 8^q on their left sides and slopes -8^q on their right sides.

In the sum in the above identity: The right side of $\tau_{q,h-7}$ (slope $= -8^q$) is superimposed on the left side of $2 \cdot \tau_{q,h-6}$ (slope $= 2 \cdot 8^q$), giving a resultant slope of 8^q. Similarly, the right side of $2 \cdot \tau_{q,h-6}$ (slope $= -2 \cdot 8^q$) is superimposed on the left side of $3 \cdot \tau_{q,h-5}$ (slope $= 3 \cdot 8^q$), giving the same resultant slope 8^q. And so on.

Finally, the sum is multiplied by $1/8$, giving a resultant slope of 8^{q-1}: exactly as for the "big" triangle function τ_{q-1}^*. This proves the identity. $\square$

For the trapezoidal function σ we have the identity:

$$\sigma(x) = \sum_{i=-M+1}^{M-1} \tau_{0i}(x)$$

(See Figure 2.)

The proof of this identity is similar to the previous proof (and easier), and we leave it to the reader. $\square$

We conclude this subsection with two inequalities derived from the above identities. To set the stage, we recall that, for $q \geq 1$, $\tau_{q-1}^* = \tau_{q-1,j}$ decomposes into a linear combination of the fifteen functions τ_{qi}, $h - 7 \leq i \leq h + 7$ ($h = 8j$).

Now fix a vector x_n from the sequence $\{x_n\}$ constructed above. Following the operational calculus of Pre-step A, we are interested in the norms $\|\tau_{qi}(T)(x_n)\|$.

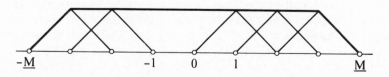

Figure 2

2. Preliminaries

We define:

u is the first index, $h - 7 \leq u \leq h + 7$ ($h = 8j$),

which maximizes $\|\tau_{qu}(T)(x_n)\|$.

Then for this u we have:

$$\|\tau_{qu}(T)(x_n)\| \geq (1/8)\|\tau_{q-1}^*(T)(x_n)\| \quad (q \geq 1),$$
$$\|\tau_{0u}(T)(x_n)\| \geq 1/(2M - 1).$$

Proofs. For the first inequality. This follows immediately from the first identity, $\tau_{q-1}^* = (1/8)[\tau_{q,h-7} + 2 \cdot \tau_{q,h-6} + \cdots]$, upon observing that the sum of the coefficients $(1/8)[1 + 2 + 3 + \cdots + 7 + 8 + 7 + \cdots + 3 + 2 + 1]$ is equal to 8.

For the second inequality. First we recall that $\|x_n\| \geq 1$, and that since $\sigma(x) = 1$ on spectrum(T), $\sigma(T)$ = identity operator. Hence $\|\sigma(T)(x_n)\| = \|x_n\| \geq 1$. Now the second inequality follows at once from the second identity, $\sigma = \sum_{-M+1}^{M-1} \tau_{0i}$, upon observing that the sum has $(2M - 1)$ terms. $\square$

Note. We do not claim that the maximizing index u can be found effectively. In the formal proof in Section 5, the index u will be replaced by a slightly inferior index v which is computed effectively.

Pre-step C (the computed norms)

To compute norms, we have to compute real numbers. Of course, a computable real number is the effective limit of a computable sequence of rationals. Thus when we "compute" a real number, the things which we actually compute are rational approximations.

We begin with the final corollary in Pre-step A, which tells us that $\{\|f_m(T)(y_n)\|\}$ is computable for any computable sequence of continuous functions $\{f_m\}$ and any computable sequence of vectors $\{y_n\}$. For $\{y_n\}$, we take the sequence of vectors $\{x_n\}$ constructed prior to Pre-step A. For $\{f_m\}$ we take the double sequence of triangle functions $\{\tau_{qi}\}$ constructed in Pre-step B. Hence we have:

$$\{\|\tau_{qi}(T)(x_n)\|\}$$

is a computable triple sequence of real numbers. We emphasize that this "computability" is simultaneously effective in all three variables, q, i and n.

Thus there exists a computable triple sequence of *rational* approximations, which we denote by $\text{CompNorm}_{qi}(n)$ such that:

$$|\text{CompNorm}_{qi}(n) - \|\tau_{qi}(T)(x_n)\|| \leq (1/1000)(1/2M)(1/16^q) \quad (q \geq 1),$$

$$|\text{CompNorm}_{0i}(n) - \|\tau_{0i}(T)(x_n)\|| \leq \frac{1}{1000}\left[\frac{1}{2M-1} - \frac{1}{2M}\right].$$

Note. Sometimes, in situations where the variable n is being temporarily held fixed, we shall write CompNorm_{qi} in place of $\text{CompNorm}_{qi}(n)$.

3. Heuristics

The two subsections Heuristics I and II below treat respectively: I. the construction itself and II. the proof of its properties.

In this heuristic section, we shall make one simplification. As a result, the "construction" described here is not effective: it contains one non-effective step. (We will flag the place where this occurs.) Later, in Section 4, we give an effective construction, followed in Section 5 by detailed proofs.

Heuristics I. A simplified version of the procedure

We now expand upon some comments made in the Introduction.

In order to motivate the steps which follow, it is useful to return momentarily to the spectral theorem in its traditional (noneffective) setting. We recall that, associated with any bounded self-adjoint operator T, there is a "spectral measure". (For details, cf. Section 1.) The spectral measure gives a decomposition of the Hilbert space H into orthogonal subspaces H_I.

Let us now make this decomposition explicit in the most obvious (albeit non-effective) way. We begin by taking an interval $(-M, M]$ containing the spectrum of T. Next we partition $(-M, M]$ into "thin" subintervals $I_i = (a_{i-1}, a_i]$ in the usual fashion

$$-M = a_0 < a_1 < \cdots < a_N = M,$$

$$a_i - a_{i-1} < \varepsilon \qquad \text{for all } i.$$

Then the corresponding subspaces,

$$H_i = H_{(a_{i-1}, a_i]},$$

are orthogonal and invariant under T, and the elements of H_i are "approximate eigenvectors" (i.e. $\|Tx - a_i x\| \leq \varepsilon \|x\|$ for $x \in H_i$). Thus we obtain a rough "picture" of the operator T, a picture that becomes more precise as we let $\varepsilon \to 0$.

Of course, these steps are wildly nonconstructive. Indeed, even ignoring the Hilbert space aspects, the question of whether a real number t belongs to an interval $(a_{i-1}, a_i]$ cannot be decided effectively. Thus we must find an analog of this procedure—one which has some chance of being effective. To achieve this, we shall have to abandon the "natural" decomposition of the real line into disjoint subintervals.

Our modification of the "disjoint interval" procedure involves two main steps. Firstly, we replace the disjoint intervals by intervals which overlap. Secondly, we eliminate the intervals altogether! More precisely, instead of considering the

3. Heuristics

characteristic functions of these intervals (step functions) we use triangle functions. The necessary triangle functions were introduced in Pre-step B above. Their advantage over step functions lies in the fact that they are continuous. This, as we shall see, is what makes effective computation of the spectrum possible.

Now we turn to the details of the "construction". This "construction" is (if we ignore its one noneffective step) a universal procedure which begins with any computable vector x and produces a computable real number λ. Likewise, if we input a computable sequence of vectors $\{x_n\}$, it produces a computable sequence of reals $\{\lambda_n\}$. In fact we shall input the computable sequence $\{x_n\}$ defined in Section 2. The resulting sequence $\{\lambda_n\}$ will be the computable sequence of reals whose existence was asserted in the Second Main Theorem.

We shall describe the procedure for a single vector x_n, but in a manner which is clearly effective in n. Then to deal with the entire sequence $\{x_n\}$, we merely use an effective process which returns to each x_n infinitely often.

Thus we *fix* a vector x_n from the computable sequence $\{x_n\}$ given in Section 2. The following procedure will lead to the corresponding real number λ_n.

Step 1. We recall the computable double sequence of triangle functions τ_{qi} defined in Pre-step B of Section 2. The function τ_{qi} is supported on the interval

$$I_{qi} = [(i-1)8^{-q}, (i+1)8^{-q}]$$

which has half-width $= 8^{-q}$. These intervals overlap in the manner

$$\ldots [-2, 0], [-1, 1], [0, 2], [1, 3], \ldots .$$

Step 2. We now have, by Pre-step A, that the double sequence of vectors $\{\tau_{qi}(T)(x_n)\}$ (recall that x_n is fixed) and the double sequence of norms $\{\|\tau_{qi}(T)(x_n)\|\}$ are computable. The norms, of course, are a computable double sequence of nonnegative real numbers.

Step 3. For each $q = 0, 1, 2, \ldots$, we shall choose an index $i = i(q)$ in a manner to be described below. This will yield a nested sequence of intervals

$$I_q^* = I_{q, i(q)}$$

where

$$I_0^* \supseteq I_1^* \supseteq I_2^* \supseteq \cdots$$

Of course, the sequence $\{I_q^*\}$ is defined by induction. To obtain the nested intervals we do the following. At any stage $q \geq 1$, *we consider only those i such that $I_{qi} \subseteq I_{q-1}^*$.*

We recall from Pre-step B that, if we write $j = i(q-1)$ so that $I_{q-1}^* = I_{q-1,j}$, then the allowed values for $i = i(q)$ must come from the finite list:

$$i = 8j - 7, \ldots, 8j + 7.$$

Finally, we observe that, corresponding to the intervals $I_q^* I_{q, i(q)}$, there is a sequence of triangle functions $\tau_q^* = \tau_{q, i(q)}$.

Step 4. We now define the sequence of intervals $\{I_q^*\}$ and triangle functions $\{\tau_q^*\}$ by induction on q. We begin with the vacuous case $q = -1$ (for which there are no triangle functions and no interval I_q^*). Then, subject to the restriction from Step 3,

$$I_{qi} \subseteq I_{q-1}^* \quad \text{(vacuous when } q - 1 = -1\text{)},$$

we choose the value $i = i(q)$ which

$$\textit{maximizes } \|\tau_{qi}(T)(x_n)\|.$$

In case of ties, we choose the smallest tying i.

[This, of course, is the noneffective step! For the norms $\|\tau_{qi}(T)(x_n)\|$ are computable real numbers, and exact comparisons between computable reals cannot be made effectively.]

Step 5 (Definition of λ_n). We define the real number λ_n as the common intersection point of the intervals I_q^* ($= I_q^*(n)$, where we have suppressed the variable n). Since the q-th interval I_q^* has half-width 8^{-q}, the convergence of these intervals as $q \to \infty$ is effective in both q and n.

Notes. Later, in Section 4, we shall obtain an effective procedure by replacing the norms $\|\tau_{qi}(T)(x_n)\|$ in Step 4 by the approximations CompNorm_{qi} from Pre-step C. Since the values $\{\text{CompNorm}_{qi}\}$ form a computable double sequence of *rationals*, exact comparisons of the CompNorm_{qi} can be made effectively.

On the other hand, the use of approximate values complexifies the proof to a substantial degree. Furthermore, the key ideas of the proof lie elsewhere. That is why, in this heuristic section, we ignore this painful but necessary step.

Not an eigenvalue!

Now we must define the set A of indices such that the set of eigenvalues coincides with $\{\lambda_n : n \notin A\}$. Thus A is to be a recursively enumerable set of natural numbers, whose significance is the following: when an integer n appears on the list A, then λ_n will *not* be counted as an eigenvalue. Thus the statement that $n \in A$ corresponds to the declaration "Not an eigenvalue!" for λ_n.

[We remark, however, that the sequence $\{\lambda_n\}$ need not be one to one. The same real number λ may appear as the value λ_n for several n. In fact, we can have $\lambda = \lambda_n = \lambda_m$ with $m \neq n$ and the declaration "Not an eigenvalue!" could be made for n but not for m. More on this below.]

The idea behind our definition of the set A is embodied in the following two facts:
(i) If λ_n is not an eigenvalue for T, then the norms $\|\tau_q^*(T)(x_n)\| \to 0$ as $q \to \infty$.
(ii) If λ_n is an eigenvalue, AND if x_n is "sufficiently close" to the corresponding eigenvector, then the norms $\|\tau_q^*(T)(x_n)\|$ remain bounded away from zero as $q \to \infty$. In fact, $\|\tau_q^*(T)(x_n)\| \geq 1/8$ for all q.

3. Heuristics

We shall prove these statements in subsequent sections. Accepting their truth for now, we can see at once what the criterion for "Not an eigenvalue!" should be.

Definition of set A ("Not an eigenvalue!"). We say that $n \in A$ if, for some q, the norm $\|\tau_q^*(T)(x_n)\| < 1/8$.

We conclude this descriptive section with several remarks.

First, it is clear that the set A defined above is recursively enumerable. For we can compute $\|\tau_q^*(T)(x_n)\|$ for $q = 0, 1, 2, \ldots$, and if a value of q with $\|\tau_q^*(T)(x_n)\| < 1/8$ ever occurs, we will eventually find it. However, there may be no effective procedure for listing the complement of A. For it is, of course, impossible to scan the entire sequence $\{\|\tau_q^*(T)(x_n)\|\}$ in a finite number of steps.

What does this mean from the viewpoint of spectral theory? In a deliberately vague but suggestive fashion, we can describe the situation as follows. We recall from Section 1 that λ is an eigenvalue if and only if there is a nonzero spectral measure concentrated in the point-set $\{\lambda\}$. Now the sequence of triangle functions $\{\tau_{qi}\}$ gives us a kind of "microscope" which allows us to examine intervals on the real line, locating those intervals where the spectral measure is most heavily concentrated. However, for each fixed q, the microscope has only a limited amount of resolving power. This power increases towards infinity as $q \to \infty$. Nevertheless, at any finite stage, our imperfect microscope is incapable of distinguishing between a single spectral line and a thin band of continuous spectrum. And, since in any effective process we are always at some finite stage, this difficulty can never by resolved. This explains heuristically why, in the case where λ_n actually is an eigenvalue, we may never possess an effective verification of that fact.

Remark. We have noted that the sequence $\{\lambda_n\}$ need not be one to one. This has the consequence that, even if λ_n is an eigenvalue, the declaration "Not an eigenvalue!" might be made for n. This is because (ii), on which the definition of A was based, requires that λ_n be an eigenvalue *and* that x_n be an approximate eigenvector. Even if λ_n satisfies this, x_n might not. However, if x_n fails, then a suitable vector x_m will eventually turn up. The new vector will give the same value $\lambda = \lambda_n = \lambda_m$, but this time the pair (λ_m, x_m) will pass the "eigenvalue/eigenvector test". For a proof of these statements, see proposition 4, whose proof is sketched in Heuristics II and then done carefully in Section 5.

Finally, we repeat that the "almost effective procedure" in this section is only an approximation to the effective (but more complicated) algorithm given in Section 4.

Heuristics II. Why the Procedure Works

In order to satisfy the conditions of the Second Main Theorem, there are four things that we must show about the sequence $\{\lambda_n\}$ and the set A "constructed" in Heuristics I. These are:

1. Every $\lambda_n \in \text{spectrum}(T)$.
2. The sequence $\{\lambda_n\}$ is dense in spectrum(T).
3. If λ_n is not an eigenvalue of T, then $n \in A$ (i.e. the declaration "Not an eigenvalue!" is made for n).
4. If λ is an eigenvalue of T, then there exists some $n \notin A$ such that $\lambda = \lambda_n$.

[Propositions 1. and 2. are virtually identical to the statements about $\{\lambda_n\}$ and spectrum(T) made in (i) of the Second Main Theorem. Propositions 3. and 4. combine to give the result: The set of eigenvalues of T coincides with the set $\{\lambda_n : n \notin A\}$—exactly as in (ii) of the Second Main Theorem. As stated earlier, (iii) and (iv) will be proved in Section 8.]

Precise proofs of 1., 2., 3., and 4. will be given in Section 5. Here, instead of trying to be semi-precise, we shall be rather casual. Yet the sketchy "proofs" which we outline here already contain the key ideas of the detailed proofs which are to come.

Consider the "construction" in Heuristics I. Recall that it begins with a vector x_n and ends with a corresponding real number λ_n. As a first step, we must unravel the meaning of this construction in terms of the Spectral Theorem.

[This will require a rather lengthy discussion. We cannot avoid it. The Spectral Theorem involves three different structures: projections/subspaces/measures, and it is the interplay between these that is vital to our construction.]

Recall that in the Spectral Theorem there are two types of spectral measures. Firstly, there is the spectral measure associated with the operator T. It consists of projections onto subspaces of the Hilbert space. Secondly, there are the measures $d\mu_{xx} = d\mu(x)$ associated with a vector x. These are ordinary positive real-valued measures.

Here we recall some notation from Section 1: $d\mu(x)$ is a measure on the real line, determined by the vector x. In integration formulas, we may want to display the real variable t (for example, as in $(T^2(x), x) = \int_{\mathbb{R}} t^2 \, d\mu(x, t)$). So we sometimes add the variable t and write $d\mu(x) = d\mu(x, t)$.

The measure $d\mu(x) = d\mu(x, t)$ is governed by the defining equation:

$$(f(T)(x), x) = \int_{-\infty}^{\infty} f(t) \, d\mu(x, t),$$

where f is any bounded Borel function, and $f(T)$ is the corresponding operator.

Now we must recall the connection between the spectral measure of T and that of x. Begin with T. For each interval $(a, b]$ in the real line, the spectral measure of T gives a projection $E_{(a,b]}$ onto a subspace $H_{(a,b]}$ of the Hilbert space H.

As we showed in Section 1 (SpThm 6), the connection between these projections and the measure $d\mu(x) = d\mu(x, t)$ is:

Let $x_0 = E_{(a,b]}(x)$ be the projection of x on the subspace $H_{(a,b]}$. Then

$$\|x_0\|^2 = \text{the } d\mu(x)\text{-measure of } (a, b].$$

In particular,

$$\|x\|^2 = \text{the } d\mu(x)\text{-measure of } \mathbb{R}.$$

As an easy application of SpThm 6, we obtain a continuous analog of the Pythagorean Theorem. Suppose we partition $\mathbb{R}$ into countably many disjoint inter-

3. Heuristics

vals $(a_i, b_i]$. Let

$$x_i = E_{(a_i, b_i]}(x),$$

so that x_i is the projection of x onto the subspace $H_{(a_i, b_i]}$. Then:

$$\sum_i \|x_i\|^2 = \|x\|^2.$$

Here is the proof. The intervals $(a_i, b_i]$ are disjoint and their union is $\mathbb{R}$. Then the fact that $d\mu(x)$ is a measure (i.e. countably additive) means that the $d\mu(x)$-measures of the $(a_i, b_i]$ add up to the $d\mu(x)$-measure of $\mathbb{R}$. But the $d\mu(x)$-measure of $(a_i, b_i]$ is $\|x_i\|^2$, and the $d\mu(x)$-measure of $\mathbb{R}$ is $\|x\|^2$.   q.e.d.

Thus we reach a conclusion which can be put into words as follows:

The measure $d\mu(x)$ shows the way the vector x breaks down into its orthogonal components x_i—while in a parallel fashion the Hilbert space is being broken down into orthogonal subspaces by the action of the projection-valued measure dE. More precisely, $d\mu(x)$ records the way that the square-norms $\|x_i\|^2$ add up via the "Pythagorean Theorem" to give the square-norm of x.

This completes our review of spectral measures.

We apply this now to the vector x_n with which we began the construction in Heuristics I. The spectral measure of x_n would seem to be a difficult thing to get our hands on computationally. Let us worry about that later. For now, let us suppose that we can somehow "see" the spectral measure, as though it were displayed on a screen like a computer-graphic.

We know that the spectral measure of x_n describes the connection between intervals on the real line and the orthogonal decomposition of the vector x_n. ("The Pythagorean Theorem"). Thus, for example, consider the interval $(a, b]$ and its associated subspace $H_{(a,b]}$: if $x_n \in H_{(a,b]}$, then all of the spectral measure of x_n is contained in $(a, b]$. If x_n deviates only slightly from a vector in $H_{(a,b]}$, then *most* of the spectral measure of x_n will lie within $(a, b]$. Of course, for most vectors x_n, the spectral measure of x_n is spread out all over the place. But, even for such x_n as these, there should be parts of the real line where the spectral measure of x_n is "more heavily concentrated". We should be able to find these regions of "heavy concentration" by partitioning the real line and choosing the subintervals which have the "heaviest concentration".

All of this, we recall, was under the assumption that we could somehow "see" the spectral measure of x_n. But the triangle functions τ_{qi} allow us to do precisely that. Our procedure in Heuristics I, of choosing the i which maximizes $\|\tau_{qi}(T)(x_n)\|$, allows us to pick out a nested sequence of subintervals of "heavy concentration" converging down to a point λ_n of "heavy concentration".

Perhaps it is not clear at a glance that the triangle functions do what we have just claimed for them. What connection is there between the norms $\|\tau_{qi}(T)(x_n)\|$ and the spectral measure $d\mu(x_n, t)$ of x_n? Well,

$$\|\tau_{qi}(T)(x_n)\|^2 = (\tau_{qi}(T)(x_n), \tau_{qi}(T)(x_n)) = ((\tau_{qi})^2(T)(x_n), x_n),$$

and by the defining equation of the spectral measure $d\mu(x, t)$ (see above), this becomes

$$\int_{-\infty}^{\infty} (\tau_{qi})^2(t) \, d\mu(x_n, t).$$

Thus the value $\|\tau_{qi}(T)(x_n)\|^2$ is equal to that gotten by integrating the square of the triangle function, $(\tau_{qi})^2$, against the spectral measure $d\mu(x_n, t)$.

[The fact that the square-norm $\|\tau_{qi}(T)(x_n)\|^2$ is the integral of the square $(\tau_{qi})^2$ is, of course, another variant of the "Pythagorean Theorem".]

Now we return to the procedure in Heuristics I. Recall that, at stage q, it involves letting i vary and choosing the i which maximizes $\|\tau_{qi}(T)(x_n)\|$. We observe that, for fixed q and varying i, the functions τ_{qi} all have the same shape—they are merely translates of one another. So the integral of $(\tau_{qi})^2$ against $d\mu(x_n, t)$ will be maximal only when the support of the function τ_{qi} contains "its fair share" of the measure $d\mu(x_n, t)$. The procedure is pushing us towards a place λ_n on the real line where the spectral measure of x_n is "heavily concentrated".

The results of the discussion—which we state in a deliberately vague but intuitively suggestive fashion—are:

The norms $\|\tau_{qi}(T)(x_n)\|$ give us a computationally effective way to get our hands on the spectral measure of x_n. Using these norms as in Heuristics I, we have a procedure which passes from an input vector x_n to an output number λ_n. This procedure is designed so that λ_n lies at a place on the real line where the spectral measure of x_n is "heavily concentrated".

Now we ask: what does this say about the propositions 1., 2., 3., 4., listed above. Let us go through them in order. For convenience, each proposition has been restated (in parentheses) at the beginning of its paragraph.

[Incidentally, the fact that in the above sentence, the phrase "in parentheses" is in parentheses, does not lead to a new type of self-referential formula in logic.]

This is a good place to pause. We have reached a turning point. The heuristic descriptions have come to an end, and the time to apply them has begun. We reemphasize that the arguments given below are intended as proof sketches, and not as formal proofs. For that reason we have omitted the usual symbol □ which marks the end of a proof. The formal proofs will be given in Section 5.

Remark concerning SpThm N. In Section 1 we gave seven results, SpThm 1 to SpThm 7, on which this proof would be based. SpThm 6 and 7 have already been used. The other five results are used in the arguments below.

1. ($\lambda_n \in \text{spectrum}(T)$.) Obviously, if λ_n lies at a point where the spectral measure is "heavily concentrated", it must lie in the spectrum. For the complement of spectrum(T) is an open set containing no spectral measure whatsoever.

2. ($\{\lambda_n\}$ *is dense in* spectrum(T).) Take any $\lambda \in \text{spectrum}(T)$. We have to show that there exist λ_n lying arbitrarily close to λ. Now the position of λ_n on the real line depends on the initial vector x_n. Obviously we must pick x_n correctly. Well, suppose we take a closed ε-neighborhood $[\lambda - \varepsilon, \lambda + \varepsilon]$ of λ. By SpThm 3 in Section 1, this neighborhood has nonzero spectral measure. Choose a vector x_n whose spectral

3. Heuristics

measure lies entirely within $[\lambda - \varepsilon, \lambda + \varepsilon]$. (This can only be done approximately—a difficulty we ignore for now.)

Let λ_n be the number corresponding to x_n. We want to show that $\lambda_n \in [\lambda - \varepsilon, \lambda + \varepsilon]$. Suppose otherwise. We recall that the point λ_n is the intersection, as $q \to \infty$, of support(τ_q^*). Then as $q \to \infty$, support(τ_q^*) approaches λ_n, and hence becomes disjoint from $[\lambda - \varepsilon, \lambda + \varepsilon]$. But $[\lambda - \varepsilon, \lambda + \varepsilon]$ contains the entire spectral measure of x_n. So by SpThm 1, for all sufficiently large q, $\|\tau_q^*(T)(x_n)\| = 0$.

However, our construction guarantees that the triangle functions τ_q^* are supported on an interval where the spectral measure of x_n is "heavily concentrated". Hence, in particular, the vector $\tau_q^*(T)(x_n) \neq 0$, and the norm $\|\tau_q^*(T)(x_n)\| > 0$. This contradicts the conclusion of the preceding paragraph.

3. (*If λ_n is not an eigenvalue, then $n \in A$: that is, the declaration "Not an eigenvalue!" will be made for n.*) Here it turns out that the spectral measure of T—involving projections onto subspaces—is easier to use than the spectral measure of x_n. We saw in Section 1 that a number λ is an eigenvalue of T if and only if the point-set $\{\lambda\}$ has nonzero spectral measure: i.e. if and only if the subspace $H_{\{\lambda\}}$ is nonzero (SpThm 5).

Suppose that λ_n is not an eigenvalue. Then the point-set $\{\lambda_n\}$ has zero spectral measure. Hence, by the countable additivity of spectral measure, the spectral measure of the interval $(\lambda_n - \varepsilon, \lambda_n + \varepsilon)$ shrinks to zero as $\varepsilon \to 0$. Now the triangle functions τ_q^* are uniformly bounded ($0 \leq \tau_q^*(t) \leq 1$) and their supports shrink to the point-set $\{\lambda_n\}$ as $q \to \infty$. Hence for any vector x_n, the vectors $\tau_q^*(T)(x_n) \to 0$, and thus the norms $\|\tau_q^*(T)(x_n)\| \to 0$ (SpThm 2). Since these norms approach zero as $q \to \infty$, they must eventually drop below the cut-off value of $1/8$. When that happens, the declaration "Not an eigenvalue!" will be made. This finishes proposition 3.

Before coming to 4., there is a caution which we must stress. The converse of 3. is false. That is, even if the declaration "Not an eigenvalue!" is made, it could still happen that the value $\lambda = \lambda_n$ is an eigenvalue. This was discussed in the Remark at the end of Heuristics I. Here we merely recall that the sequence $\{\lambda_n\}$ need not be one to one. Therefore if $\lambda_n = \lambda_m = \lambda$ with $m \neq n$, we might have the "Not an eigenvalue!" declaration for n but not for m.

4. (*If λ is an eigenvalue of T, then there exists an $n \notin A$ with $\lambda = \lambda_n$.*) As in 2. above, the trick is to choose x_n correctly. We begin with an x_n which is an eigenvector of T with eigenvalue λ. (Again this can only be done approximately, a difficulty we ignore for now.) Since x_n is an eigenvector for λ, spectral theory (SpThm 4) tells us that, for any continuous function f,

$$f(T)(x_n) = f(\lambda) \cdot x_n.$$

In particular, this holds for the triangle functions τ_{qi}:

$$\tau_{qi}(T)(x_n) = \tau_{qi}(\lambda) \cdot x_n.$$

Assume, for convenience, that x_n is a unit vector (again only approximately true). Then:

$$\|\tau_{qi}(T)(x_n)\| = \tau_{qi}(\lambda).$$

Now $\tau_{qi}(\lambda)$ is a much more tractable thing to deal with than $\|\tau_{qi}(T)(x_n)\|$. We can simply look at the graphs of the triangle functions τ_{qi} and see the way they overlap (see Figures 1 and 2 in Section 2 above). We see from the graphs of the τ_{qi} that:

For any q, there exists an i, such that $\tau_{qi}(\lambda) \geqslant 1/2$.

Recall that τ_q^* denotes the function τ_{qi} which maximizes $\|\tau_{qi}(T)(x_n)\| = \tau_{qi}(\lambda)$. Thus we have:

$$\tau_q^*(\lambda) \geqslant 1/2 \qquad \text{for all } q.$$

Recall further that, by definition, λ_n is the common intersection of the intervals $I_q^* = \text{support}(\tau_q^*)$.

The results which we need to show are (i) $\lambda = \lambda_n$, and (ii) $n \notin A$.

(i) $\lambda = \lambda_n$. Well, $\tau_q^*(\lambda) \geqslant 1/2$ for all q, which puts λ within the support of τ_q^* for all q. Hence $\lambda \in I_q^*$ for all q, $\lambda_n \in I_q^*$ for all q (by definition), and since the widths of the intervals I_q^* shrink to zero, $\lambda = \lambda_n$.

(ii) $n \notin A$ (i.e. the declaration "Not an eigenvalue!" is never made). Well, this declaration will be made if, for some q, $\|\tau_q^*(T)(x_n)\| = \tau_q^*(\lambda) < 1/8$. But we have seen that $\tau_q^*(\lambda)$ remains always $\geqslant 1/2$.

Note. The reader may wonder why we chose the value 1/8 as our cut-off for the "Not an eigenvalue!" declaration. Actually, any value strictly less than 1/2 would do. However, for normal operators (c.f. Section 6), we need a value $< 1/4$. We wanted a uniform procedure, so we simply took the next power of two below 1/4.

This completes our discussion of proposition 4.

We make one final comment. Throughout this section, we have been rather cavalier about "approximations". A computable real number can only be known approximately. For the most part, this is a mere nuisance. However, there is one place in our construction where the need to approximate plays a pivotal role. This is the fact that our triangle functions have overlapping supports, after the manner of

$$\ldots [-2, 0], [-1, 1], [0, 2] \ldots .$$

Why do we do this? Suppose instead that we used intervals which abut, like

$$\ldots [-1, 0], [0, 1], [1, 2] \ldots .$$

What would go wrong? The trouble is that something like the following might happen:

There might be an eigenvalue λ which is slightly greater than 1, but which due to errors in computation we reckon as being slightly less than 1. Consequently we select the interval $[0, 1]$ instead of $[1, 2]$. Well …? The eigenvalue λ has been lost forever. For our method requires us to remain within the interval $[0, 1]$. No amount of partitioning of $[0, 1]$ can bring us back to λ, which lies outside of $[0, 1]$.

By using overlapping intervals we avoid these difficulties.

4. The Algorithm

In this section we give the algorithm and prove that it is effective.

We recall that there are two constructions which we must carry out. We must construct the sequence $\{\lambda_n\}$ of real numbers, and list the set A of indices for which the declaration "Not an eigenvalue!" is made. We repeat that the algorithm below is NOT the same as the "construction" given in Heuristics I. The procedure in Heuristics I was a simplified (noneffective) preview of what we do here.

We shall give this algorithm in the form of a "recipe", simply listing its steps. Any explanations we include will be of a descriptive nature (to clarify what the recipe *is*). As already noted, the properties of this algorithm are proved in Section 5.

Construction of the λ_n. The number λ_n will be constructed via a universal effective process applied to the vector x_n. We shall describe this process for a single *fixed* value of n. The sequence $\{\lambda_n\}$ is then generated by using a recursive procedure which returns to each n infinitely often.

The ingredients for this construction are:

The vector x_n, which is held fixed.
The operational calculus of Pre-step A.
The triangle functions τ_{qi} of Pre-step B.
The approximations CompNorm_{qi} of Pre-step C.

We recall that the triangle functions τ_{qi} are supported on the intervals $I_{qi} = [(i-1)8^{-q}, (i+1)8^{-q}]$ of half-width 8^{-q}. These intervals overlap in the manner:

$$\ldots [-2, 0], [-1, 1], [0, 2], [1, 3] \ldots.$$

[The trapezoidal function σ plays no role in the construction, although it will play a role in the proofs which follow.]

We recall further that the values CompNorm_{qi} are a computable multi-sequence of nonnegative *rational* numbers which approximate the norms $\|\tau_{qi}(T)(x_n)\|$ to within an error given by:

$$|\text{CompNorm}_{qi} - \|\tau_{qi}(T)(x_n)\|| \leq (1/1000)(1/2M)(1/16^q).$$

Now here is the recipe:

We proceed by induction, beginning with the stage $q = -1$, and going forward to the stages $q = 0, 1, 2, \ldots$. At the stage $q = -1$, nothing has been done.

At each stage $q \geq 0$ we shall select a single triangle function τ_q^* from among the τ_{qi}, $-M \cdot 8^q < i < M \cdot 8^q$, defined in Pre-step B. If one wants to be very formal, we are really selecting an index $i = i(q)$ from the list of indices $-M \cdot 8^q < i < M \cdot 8^q$. Then we have the triangle τ_q^*, the interval I_q^*, and the computed norm CompNorm_q^* given by:

$$\tau_q^* = \tau_{qi}, \qquad i = i(q);$$
$$I_q^* = I_{qi}, \qquad i = i(q);$$
$$\text{CompNorm}_q^* = \text{CompNorm}_{qi}, \qquad i = i(q).$$

Now, at any stage $q \geq 1$, we impose the following restriction. In selecting $\tau_q^* = \tau_{qi}$ [i.e. in choosing $i = i(q)$], *we consider only those i such that*

$$I_{qi} \subseteq I_{q-1}^*.$$

[This guarantees that the selected interval I_q^* satisfies $I_q^* \subseteq I_{q-1}^*$. Hence $I_0, I_1, I_2, \ldots$ form a nested sequence of intervals.]

It may be useful to recall (cf. Pre-step B) that, if $I_{q-1}^* = I_{q-1,j}$, then the allowed values for $i = i(q)$ are

$$i = 8j - 7, \ldots, 8j + 7.$$

It is from this finite list that the actual value $i = i(q)$ will be selected.

Finally, we are ready to describe the selection process for $i = i(q)$. It is this. Subject to the above restrictions on i:

Choose the i for which CompNorm_{qi} is maximal.

In case of ties, we take the smallest tying i.

This is an effective process, since the multi-sequence CompNorm_{qi} is a computable double sequence of *rational* numbers.

The number λ_n is defined as the common intersection of the intervals I_q^* ($= I_{qi}$ for $i = i(q)$).

Lemma. *The sequence $\{\lambda_n\}$ is computable.*

Proof. The I_q^* form a computable nested sequence of intervals of half-widths 8^{-q}. Clearly these half-widths approach zero effectively. Hence the above furnishes an effective procedure for computing the real number λ_n.

What about effectiveness in n? For convenience in description, we have held n fixed. But, clearly, the procedure is effective in n also. For $\{x_n\}$ is a computable sequence of vectors, and the procedures in Pre-steps A and B are canonical and universal. By contrast, the approximation procedure in Pre-step C—giving the rational approximations CompNorm_{qi}—is slightly less canonical. But we took pains to insure that the computation was effective in n as well as q and i. That is all we need. □

This completes the construction of $\{\lambda_n\}$.

Construction of the Set A (Not an eigenvalue!). Recall that A is to be a recursively enumerable set of natural numbers such that the set of eigenvalues of T has the form $\{\lambda_n : n \notin A\}$. Roughly speaking, A gives the set of indices n for which we make the declaration "Not an eigenvalue!" The set A, although recursively enumerable, need not be recursive. Thus, from a computational point of view, we will eventually compute the $n \in A$—i.e. those n for which the declaration "Not an eigenvalue!" is made. But the complementary set, listing the sequence of eigenvalues themselves, may never be known to us.

Of course, each *individual* eigenvalue (as opposed to the sequence of all eigenvalues) is computable. For if we hold n fixed, then the set A becomes irrelevant. We simply fix n and then apply the effective procedure above for computing λ_n.

We turn now to the effective listing of the set A. Continuing with the notation used in constructing $\{\lambda_n\}$, we shall add one detail. All of the previous constructions depended on the initial vector x_n, which we held fixed. Where previously we suppressed the variable n, now it will be useful to display it. Thus we write:

$$i(q) = i(q, n),$$
$$\tau_q^* = \tau_q^*(n),$$
$$I_q^* = I_q^*(n),$$

and in particular,

$$\text{CompNorm}_q^* = \text{CompNorm}_q^*(n).$$

Now it is easy to describe the set A:
$n \in A$ if and only if, for some $q = 0, 1, 2, \ldots,$

$$\text{CompNorm}_q^*(n) < \tfrac{1}{8}.$$

Lemma. *The set A is recursively enumerable.*

Proof. Clearly the set A (although not its complement) can be effectively listed. For, firstly, $\{\text{CompNorm}_q^*(n)\}$ is a computable double sequence of *rational* numbers. Thus, to list A, we scan the set of pairs (n, q) in a recursive manner, returning to each n infinitely often: If $\text{CompNorm}_q^*(n) < 1/8$ for some q, then we shall eventually find this q, and consequently add the integer n to the set A. □

5. *Proof That the Algorithm Works*

We have now given all of the necessary constructions and proved their effectiveness. But we have not proved that these constructions fulfill the promises made in the Second Main Theorem. This is our final task.

The proof depends on several inequalities, and we shall begin by deriving these. Then we will turn to the propositions 1. to 4. discussed in Heuristics II. For convenience, we will restate each proposition before giving its proof.

There are three inequalities. Analytically, they are not difficult. However, the combinatorial situation which gives rise to them requires a bit of preface.

Suppose that in our construction we have completed stage $(q - 1)$ and are looking at stage q. The triangle function τ_{q-1}^* with support I_{q-1}^* has already been selected. We recall that, if we write $I_{q-1}^* = I_{q-1,j}$, then the next value of i must be selected from the list:

$$8j - 7 \leq i \leq 8j + 7.$$

We first consider what we would do if we could achieve perfect accuracy. Let u be an index from the above list which maximizes the norm $\|\tau_{qu}(T)(x_n)\|$. That is, u is chosen so that, with i restricted as above:

$$\|\tau_{qu}(T)(x_n)\| \geq \|\tau_{qi}(T)(x_n)\| \quad \text{for all } i.$$

(Of course, there is no effective procedure for finding u.)

Let v be the least index from the above list which maximizes CompNorm_{qv}. That is, v is chosen so that, with i restricted as above:

$$\text{CompNorm}_{qv} \geq \text{CompNorm}_{qi} \quad \text{for all } i.$$

Here, by contrast, we can compute v. In fact, v is just the value $v = i(q)$ which is used in the effective algorithm of Section 4. Hence, by the definitions of τ_q^* and CompNorm_q^* in Section 4:

$$\tau_q^* = \tau_{qv}$$

$$\text{CompNorm}_q^* = \text{CompNorm}_{qv}.$$

The use of the index v instead of u can lead to values which are slightly less than maximal. We must compare these "imperfect" values to $\|\tau_{qu}(T)(x_n)\|$, which is what we would obtain if we could do perfect computations. We repeat that the reason for these considerations is that we can find v effectively, but not u.

The key inequalities

$$|\text{CompNorm}_q^* - \|\tau_q^*(T)(x_n)\|| \leq (1/1000)(1/2M)(1/16^q). \tag{InEq 1}$$

$$\|\tau_q^*(T)(x_n)\| \geq \|\tau_{qu}(T)(x_n)\| - (2/1000)(1/2M)(1/16^q). \tag{InEq 2}$$

$$\|\tau_q^*(T)(x_n)\| \geq (1/2M)(1/16^q). \tag{InEq 3}$$

Proof of the inequalities. For InEq 1. The error estimates in Pre-step C give us, for all q and i:

$$|\text{CompNorm}_{qi} - \|\tau_{qi}(T)(x_n)\|| \leq (1/1000)(1/2M)(1/16^q).$$

Hence, in particular, this inequality holds for $i = v$. Since $\text{CompNorm}_q^* = \text{CompNorm}_{qv}$ and $\tau_q^* = \tau_{qv}$, this gives InEq 1. □

For InEq 2, we observe that similarly CompNorm_{qu} deviates from $\|\tau_{qu}(T)(x_n)\|$ by less than $(1/1000)(1/2M)(1/16^q)$. Now by definition of $v = i(q)$, $\text{CompNorm}_{qv} \geq \text{CompNorm}_{qu}$. (With the CompNorms, which are effectively computed rational

5. Proof That the Algorithm Works

numbers, we do, of course, pick the best value.) Hence:

$$\text{CompNorm}_{qv} \geqslant \|\tau_{qu}(T)(x_n)\| - (1/1000)(1/2M)(1/16^q).$$

Again we recall that $\text{CompNorm}_q^* = \text{CompNorm}_{qv}$. By InEq 1, the transition back from CompNorm_q^* to $\|\tau_q^*(T)(x_n)\|$ introduces another error of $(1/1000)(1/2M)(1/16^q)$. This, added to the identical error above, produces the $(2/1000)(1/2M)(1/16^q)$ of InEq 2. □

For InEq 3. We use induction on q. Assume that the inequality has been proved for $q - 1$. First we will deal with the case $q \geqslant 1$, and then we shall come back to the case $q = 0$. Take $q \geqslant 1$. By the induction hypothesis:

$$\|\tau_{q-1}^*(T)(x_n)\| \geqslant (1/2M)(1/16^{q-1}).$$

Now we use an inequality which has already been proved in Pre-step B. This inequality was based on the decomposition formulas for triangle functions (see Figures 1 and 2 above). We proved in Pre-step B that:

$$\|\tau_{qu}(T)(x_n)\| \geqslant (1/8)\|\tau_{q-1}^*(T)(x_n)\|.$$

For convenience, let us call $(1/2M)(1/16^q)$ "the target value". Our objective is to show that $\|\tau_q^*(T)(x_n)\| \geqslant$ (the target value).

By combining the two displayed inequalities above, we obtain $\|\tau_{qu}(T)(x_n)\| \geqslant 2 \cdot$ (the target value), since $(1/8)(1/16^{q-1})(1/2M) = 2 \cdot (1/16^q)(1/2M)$. We must make the transition from τ_{qu} (the true maximum) to τ_{qv} ($= \tau_q^*$, the function we select). Well, we simply use InEq 2. We have $2 \cdot$ (the target value), and the "error" in InEq 2 forces us to subtract $(2/1000) \cdot$ (the target value). This leaves us with:

$$\|\tau_q^*(T)(x_n)\| \geqslant [2 - (2/1000)] \cdot \text{(the target value)}.$$

The coefficient $[2 - (2/1000)]$ exceeds the required value of 1, with room to spare.

Now we must do the case $q = 0$. This is where the trapezoidal function σ from Pre-step B comes in. (See Figure 2.) Actually, we already did most of the work in Pre-step B, where we proved—using σ—that:

$$\|\tau_{0u}(T)(x_n)\| \geqslant 1/(2M - 1).$$

Again u is the maximizing index, v is the slightly inferior index which we select, and $\tau_0^* = \tau_{0v}$. We need $\|\tau_0^*(T)(x_n)\| \geqslant 1/2M$. Hence this allows us to use the gap between $1/(2M - 1)$ and $1/2M$.

Now we go back to Pre-step C. There, at stage $q = 0$, we insisted on an error:

$$|\text{CompNorm}_{0i} - \|\tau_{0i}(T)(x_n)\|| \leqslant \frac{1}{1000}\left[\frac{1}{2M-1} - \frac{1}{2M}\right].$$

Well, how convenient!

Again, our "error" is 1/1000-th of the allowable gap. The rest of the proof is so similar to that given above (for $q \geq 1$) that we leave any further details to the reader. □

This completes our treatment of the inequalities InEq 1 to InEq 3.

The end of the proof

We now give the proofs of the propositons 1. to 4.

Because we have made the right preparations, these proofs are very short. We recall that "SpThm N" refers to the N-th corollary of the spectral theorem, as developed in Section 1. Of course, "InEq N" refers to the N-th inequality in the preceding subsection.

1. (*Every $\lambda_n \in $ spectrum(T).*)

Proof. Suppose not. Since spectrum(T) is a closed set, λ_n must lie within an open interval $(\lambda_n - \varepsilon, \lambda_n + \varepsilon)$ outside of spectrum(T). Since the supports of the triangle functions τ_q^* shrink to the point λ_n, there must be some q for which support$(\tau_q^*) \subseteq (\lambda_n - \varepsilon, \lambda_n + \varepsilon)$.

Thus the support of τ_q^* lies entirely outside of spectrum(T). Since $\|\tau_q^*(T)\| \leq \sup\{|\tau_q^*(\lambda)|: \lambda \in \text{spectrum}(T)\}$ (SpThm 7), $\tau_q^*(T) = 0$. Hence $\|\tau_q^*(T)(x_n)\| = 0$. This contradicts InEq 3. □

2. (*The sequence $\{\lambda_n\}$ is dense in* spectrum(T).)

Proof. Let $\lambda \in $ spectrum(T), and take any $\varepsilon > 0$. We must find a λ_n such that $|\lambda_n - \lambda| < \varepsilon$.

Take an integer q such that the interval $I_q^* = $ support(τ_q^*) has length $< \varepsilon/2$, that is, such that $2 \cdot 8^{-q} < \varepsilon/2$.

From SpThm 3, the open interval $(\lambda - \varepsilon/2, \lambda + \varepsilon/2)$ corresponds to a *nonzero* subspace $H_{(\lambda-\varepsilon/2, \lambda+\varepsilon/2)}$ of H.

Let x be any unit vector in $H_{(\lambda-\varepsilon/2, \lambda+\varepsilon/2)}$.

Since $\{x_n\}$ is dense on the spherical shell $\{x: 1 \leq \|x\| \leq 1001/1000\}$, there exists some x_n with $\|x_n - x\| < (1/2M)(1/16^q)$. Thus we may write:

$$x_n = x + z,$$

$$\|z\| < (1/2M)(1/16^q),$$

$$x \in H_{(\lambda-\varepsilon/2, \lambda+\varepsilon/2)}.$$

We use this vector x_n with its associated triangle functions τ_q^* and its associated value λ_n. We recall that λ_n is the common intersection of the intervals $I_q^* = $ support(τ_q^*) for $q = 0, 1, 2, \ldots$. On the other hand, in this argument we are using a *fixed* value of q, as defined above. We claim:

$$\text{support}(\tau_q^*) \text{ intersects } (\lambda - \varepsilon/2, \lambda + \varepsilon/2).$$

5. Proof That the Algorithm Works

Suppose not. Then the support of τ_q^* lies entirely outside of $(\lambda - \varepsilon/2, \lambda + \varepsilon/2)$, whereas the vector x lies entirely within the subspace $H_{(\lambda-\varepsilon/2, \lambda+\varepsilon/2)}$. Hence by SpThm 1, $\tau_q^*(T)(x) = 0$. Hence $\tau_q^*(T)(x_n) = \tau_q^*(T)(z)$. Now $|\tau_q^*(t)| \leq 1$ for all real t, whence $\|\tau_q^*(T)\| \leq 1$ (SpThm 7), whence

$$\|\tau_q^*(T)(x_n)\| = \|\tau_q^*(T)(z)\| \leq \|z\| < (1/2M)(1/16^q).$$

This contradicts InEq 3.

Consequently $I_q^* = \text{support}(\tau_q^*)$ does intersect $(\lambda - \varepsilon/2, \lambda + \varepsilon/2)$. Since I_q^* has length $< \varepsilon/2$, and $\lambda_n \in I_q^*$, we obtain $|\lambda_n - \lambda| < \varepsilon$, as desired. □

Summary of 1. and 2. (the spectrum). This is a good place to pause and recall what went into the proofs of 1. and 2. Begin with 2. There we used a vector x_n which depended on q which in turn depended on ε. We used the inequality InEq 3 $[\|\tau_q^*(T)(x_n)\| \geq (1/2M)(1/16^q)]$, which also depends on q, but is *independent of n*. Much of the work in Pre-step B—the careful partitioning of triangles as shown in Figures 1 and 2 above—was aimed at producing this independence. It is essential. For the definition of x_n implicitly depended on InEq 3. If InEq 3 also depended on x_n, then our definition would be circular.

By contrast, the proof of 1. required only $\|\tau_q^*(T)(x_n)\| > 0$. If this were all we needed, it could be attained much more easily.

We turn now to the "eigenvalue propositions" 3. and 4.

Remarks. Here we shall not need such sharp error estimates as those proved in InEq 1 to InEq 3 above. Instead of the error term $(1/1000)(1/2M)(1/16^q)$ of InEq 1, we can get by with $1/1000$. Similarly for InEq 2. Lastly, InEq 3 has served its purpose (in proving 2.) and will not be seen again.

Of course, in this theoretical account, we are not going to alter our construction. But for the purpose of algorithmic efficiency, it might be well to record the fact: If one cared only about eigenvalues, and not about the spectrum, then a fixed "error" such as $1/1000$ would suffice.

3. (*If λ_n is not an eigenvalue, then $n \in A$, that is, the declaration "Not an eigenvalue!" is made for n.*)

Proof. Since λ_n is not an eigenvalue, the spectral measure of the point-set $\{\lambda_n\}$ is null (SpThm 5).

The triangle functions are uniformly bounded $(0 \leq \tau_q^*(t) \leq 1)$. Recall that, as $q \to \infty$, the supports of the τ_q^* shrink to the point-set $\{\lambda_n\}$. Hence, as $q \to \infty$, the functions $\tau_q^*(t) \to 0$ pointwise *except* at the point $t = \lambda_n$.

Now, as we have seen, the point-set $\{\lambda_n\}$ has spectral measure zero. Thus, as $q \to \infty$, the functions $\tau_q^*(t) \to 0$ "almost everywhere" in terms of the spectral measure. Hence by SpThm 2, $\tau_q^*(T)(x_n) \to 0$, which means by definition that $\|\tau_q^*(T)(x_n)\| \to 0$.

The rest is trivial. Since $\|\tau_q^*(T)(x_n)\| \to 0$ as $q \to \infty$, eventually $\|\tau_q^*(T)(x_n)\|$ becomes less than $(1/8) - (1/1000)$. Since $\|\tau_q^*(T)(x_n)\|$ and CompNorm$_q^*$ differ by $\leq (1/1000)$ (InEq 1), eventually CompNorm$_q^*$ becomes less than $1/8$. When that happens, the declaration "Not an eigenvalue!" is made. □

4. (*If λ is an eigenvalue of T, then there exists some $n \notin A$ with $\lambda = \lambda_n$.*)

Proof. Let λ be an eigenvalue of T, and let x be a unit eigenvector corresponding to λ. Since $\{x_n\}$ is dense on the spherical shell $\{x: 1 \leq \|x\| \leq 1001/1000\}$, there exists some x_n with $\|x_n - x\| < 1/1000$. Let λ_n be the scalar corresponding to x_n via our construction. We wish to show that $\lambda = \lambda_n$ and $n \notin A$. For this we use:

Lemma. *Under the above assumptions on x_n, we have $\tau_q^*(\lambda) \geq 1/7$ and* CompNorm$_q^* \geq 1/7$ *for all q.*

We first show that the lemma implies 4. Recall that λ_n is the common intersection point of the intervals $I_q^* = \text{support}(\tau_q^*)$ for $q = 0, 1, 2, \ldots$. Since, by the lemma, $\lambda \in \text{support}(\tau_q^*)$ for all q, $\lambda = \lambda_n$. Since CompNorm$_q^* \geq 1/7 > 1/8$, the declaration "Not an eigenvalue!" is never made, whence $n \notin A$.

Proof of lemma. We give the essential points first, and save the details of "approximation" for the end. Thus for now we work with the true eigenvector x, and ignore its approximation x_n. From SpThm 4 we have:

$$\|\tau_{qi}(T)(x)\| = \tau_{qi}(\lambda) \cdot \|x\| = \tau_{qi}(\lambda) \qquad \text{for all } q, i.$$

Now the proof hinges on the way the triangle functions τ_{qi} overlap. (Here see Figures 1 and 2 in Section 2 above, and especially Figure 3 below.) Hold λ fixed. Then for any q, there exists some index i such that $\tau_{qi}(\lambda) \geq 1/2$. However, this overlooks a crucial point. We are not allowed to pick i with complete freedom. (Here cf. the closing remarks in Heuristics II.) Specifically, the situation is this:

Suppose at stage $q - 1$ we have selected the function $\tau_{q-1}^* = \tau_{q-1, j}$. Then at stage q, the index $i = i(q)$ must come from the list

$$i = 8j - 7, \ldots, 8j + 7.$$

We must show that for one of *these* i, $\tau_{qi}(\lambda) \geq 1/2$. Here again we refer to the geometry of the triangle functions (Figure 3). One readily verifies:

(ooo) $\tau_{qi}(\lambda) \geq 1/2$ for some $i = 8j - 7, \ldots, 8j + 7$ if and only if $\tau_{q-1}^*(\lambda) \geq 1/16$—that is, if and only if λ lies within the middle 15/16-ths of the support of τ_{q-1}^*.

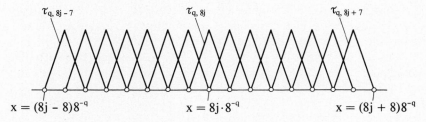

Figure 3

5. Proof That the Algorithm Works

Now the rest of the argument, while analytically simple, involves an induction on the pair of statements:

(∗) For all q, there is some $i = 8j - 7, \ldots, 8j + 7$ such that $\tau_{qi}(\lambda) \geq 1/2$.

(∗∗) $\tau_q^*(\lambda) \geq 1/7$.

Assume that both (∗) and (∗∗) hold for $q - 1$. Then by (∗∗), $\tau_{q-1}^*(\lambda) \geq 1/7$, which exceeds 1/16 with room to space. By (ooo) above, this gives (∗) [although not yet (∗∗)] for q.

We must also verify (∗) for $q = 0$. (Here see Figure 2 above.) This follows immediately from the fact that spectrum$(T) \subseteq [-(M-1), M-1]$, whence $\lambda \in [-(M-1), M-1]$.

Now we turn to the derivation of (∗∗). This is a mundane problem of approximation. Let u be the value of i, $8j - 7 \leq i \leq 8j + 7$, which maximizes $\tau_{qi}(\lambda)$. [We do not claim that u can be found effectively.] By (∗) we know that

$$\tau_{qu}(\lambda) \geq 1/2,$$

and since $\tau_{qu}(\lambda) = \|\tau_{qu}(T)(x)\|$,

$$\|\tau_{qu}(T)(x)\| \geq 1/2.$$

Since $\|x_n - x\| < 1/1000$ and $\|\tau_{qu}(T)\| \leq 1$ (SpThm 7):

$$|\|\tau_{qu}(T)(x)\| - \|\tau_{qu}(T)(x_n)\|| < 1/1000.$$

Since CompNorm$_{qu}$ differs from $\|\tau_{qu}(T)(x_n)\|$ by less than 1/1000,

$$|\|\tau_{qu}(T)(x)\| - \text{CompNorm}_{qu}| < 2/1000.$$

Since, by definition, CompNorm$_q^*$ is the maximum of the computed norms, CompNorm$_q^* \geq$ CompNorm$_{qu}$, whence

$$\text{CompNorm}_q^* \geq \|\tau_{qu}(T)(x)\| - 2/1000$$
$$\geq (1/2) - (2/1000).$$

Since CompNorm$_q^*$ differs from $\|\tau_q^*(T)(x_n)\|$ by less than 1/1000,

$$\|\tau_q^*(T)(x_n)\| \geq \|\tau_{qu}(T)(x)\| - 3/1000.$$

Finally, since $\|x_n - x\| < 1/1000$,

$$\|\tau_q^*(T)(x)\| \geq \|\tau_{qu}(T)(x)\| - 4/1000$$
$$\geq (1/2) - (4/1000).$$

Now we are back to the true eigenvector x, but with the "imperfect" triangle function τ_q^* which our approximate construction furnishes. We have (again by SpThm 4):

$$\|\tau_q^*(T)(x)\| = \tau_q^*(\lambda) \geq (1/2) - (4/1000).$$

The value $(1/2) - (4/1000)$ exceeds the desired target value of $1/7$ with room to spare. This proves (**), and completes the induction from $q - 1$ to q.

Finally, several steps back, we had $\text{CompNorm}_q^* \geq (1/2) - (2/1000)$, which exceeds $1/7$ with slightly more room to space. □

This proves the lemma. As we have seen, the lemma implies proposition 4. This, in turn, completes the proof of the Second Main Theorem. □

More precisely, we have proved the positive parts (i) and (ii) of the Second Main Theorem for the case of bounded self-adjoint operators. Normal operators, unbounded self-adjoint operators, and the negative parts (iii) and (iv) will be treated in Sections 6, 7, and 8 respectively.

6. Normal Operators

We recall that a bounded linear operator T is normal if it commutes with its adjoint, i.e. if $TT^* = T^*T$. The Second Main Theorem extends mutatis mutandis to bounded normal operators. This extension is needed for the unbounded self-adjoint case.

Theorem 1 (Normal Operators). *Let H be an effectively separable Hilbert space. Let $T: H \to H$ be a bounded normal operator. Suppose that T is effectively determined. Then there exists a computable sequence $\{\lambda_n\}$ of complex numbers, and a recursively enumerable set A of integers such that:*

>*each $\lambda_n \in \text{spectrum}(T)$;*
>
>*the spectrum of T is the closure in $\mathbb{C}$ of the set $\{\lambda_n\}$;*
>
>*the set of eigenvalues of T coincides with $\{\lambda_n : n \in \mathbb{N} - A\}$.*

Before we come to the proof of Theorem 1, we need the following.

Proposition. *Let $T: H \to H$ be an effectively determined bounded normal operator. Then the adjoint T^* is effectively determined.*

Proof. We wish to show that T^*x is computable if x is. It will be obvious that the procedure is effective, uniformly for computable sequences $\{x_k\}$.

Let x be computable. Now T is bounded and effectively determined. Hence by the First Main Theorem (Chapter 3), Tx is computable. Then by the Norm Axiom,

6. Normal Operators

$\|Tx\|$ is computable. Since T is normal, $\|Tx\| = \|T^*x\|$, for we have $\|Tx\|^2 = (T^*Tx, x) = (TT^*x, x) = \|T^*x\|^2$. Thus $\|T^*x\|$ is computable.

We recall that by Lemma 7, Section 6, Chapter 4 there exists a computable orthonormal basis $\{e_n\}$ for H. Let $\{c_n\}$ be the sequence of "Fourier coefficients" of T^*x, namely $c_n = (T^*x, e_n) = (x, Te_n)$. Since T is effectively determined, we see that $\{c_n\}$ is computable. Thus the sequence of vectors $y_n = \sum_{i=0}^{n} c_i e_i$ is computable.

The norms $\|y_n\| = \left(\sum_{i=0}^{n} |c_i|^2\right)^{1/2}$ form a nondecreasing sequence whose limit is $\left(\sum_{i=0}^{\infty} |c_i|^2\right)^{1/2} = \|T^*x\|$. As we have seen, $\|T^*x\|$ is computable. Thus the norms $\{\|y_n\|\}$ form a computable sequence converging monotonically to a computable limit $\|T^*x\|$. Hence the convergence is effective (Chapter 0, Section 2). Finally, $\|y_n - T^*x\|^2 = \sum_{i=n+1}^{\infty} |c_i|^2 = \|T^*x\|^2 - \|y_n\|^2 \to 0$ effectively—i.e. $y_n \to T^*x$ effectively in the norm of H. Since $\{y_n\}$ is computable, it follows by the Limit Axiom that T^*x is computable. This proves the proposition. □

Note. This proposition fails for operators which are not normal. That is, there exists a bounded (non-normal) effectively determined operator T whose adjoint T^* is not effectively determined. Since we do not need this counterexample, we shall not digress by presenting it.

Proof of Theorem 1

The proof follows so closely the proofs in Sections 1–5 that it is pointless to give it in detail. Instead we list the few modifications which are necessary in order to pass from the bounded self-adjoint to the normal case.

First, and obviously, we use the spectral theorem for bounded normal rather than bounded self-adjoint operators. Here the spectrum of T is a compact set in the complex plane rather than the real line. All of the (minor) modifications in the proof are consequence of this circumstance. We now list these modifications, with reference to the places in Sections 1–5 where they occur.

(Weierstrass approximation theorem in Section 2, Pre-step A). There we had real polynomials $p(t)$, where the relevant values of t were the points λ in the spectrum of T. Here we replace polynomials in the real variable λ by polynomials in the two complex variables λ and $\bar{\lambda}$. We observe that, in the operational calculus, λ corresponds to T and $\bar{\lambda}$ corresponds to T^*. For the Weierstrass theorem, we note that a arbitrary polynomial in $x = \text{Re}(\lambda)$, $y = \text{Im}(\lambda)$ can be expressed in terms of λ and $\bar{\lambda}$: $x = (\lambda + \bar{\lambda})/2$, $y = (\lambda - \bar{\lambda})/2i$.

(The two dimensional grid—cf. Section 2). As before, we partition our intervals into 8 parts at each stage. However, here the spectrum of T is complex. Hence we have a two dimensional grid on which each square is partitioned like a chessboard into 64 squares. As this happens at each stage, after q stages the number of squares is multiplied by 64^q.

(The "triangle functions" $\tau_{qi}(x)$ in Section 2, Pre-step B.) Since the spectrum of T is complex, we need functions of two real variables x, y. We set:

$$\tau_{qij}(x, y) = \tau_{qi}(x)\tau_{qj}(y), \qquad -8^q M < i < 8^q M, \qquad -8^q M < j < 8^q M.$$

In Section 2, Pre-step B we had the decomposition identity:

$$\tau_{q-1,i}(x) = \frac{1}{8} \sum_{j=-7}^{7} (8 - |j|) \cdot \tau_{q,h+j}(x), \qquad \text{where } h = 8i.$$

(As we recall, the coefficients go $1, 2, 3, \ldots, 7, 8, 7, \ldots, 3, 2, 1$.) Now by the distributive law, the product $\tau_{q-1,i}(x)\tau_{q-1,j}(y)$ decomposes into a linear combination of terms $\tau_{qr}(x)\tau_{qs}(y)$, where $8i - 7 \leq r \leq 8i + 7$, $8j - 7 \leq s \leq 8j + 7$.

For the sake of thoroughness, we check that the factor 64^q mentioned above is correct. Consider the passage from $q - 1$ to q. The function $\tau_{q-1,ij}(x, y)$ is a linear combination of products $\tau_{qr}(x)\tau_{qs}(y)$. The sum of the coefficients is

$$(\tfrac{1}{8})^2 [1 + 2 + \cdots + 7 + 8 + 7 + \cdots + 2 + 1]^2 = \tfrac{1}{64} \cdot 64^2 = 64.$$

Hence the largest of the $\|\tau_{qr}(x)\tau_{qs}(y)(T)(x_n)\|$ must be $\geq 1/64$ the size of the corresponding term for $q - 1$.

(The "CompNorms" in Section 2, Pre-step C). Previously, at stage q, the CompNorms = computed norms approximated the true norms to within an error of

$$(1/1000)(1/2M)(1/16^q) \qquad \text{for } q \geq 1.$$

Here the "$1/2M$" took care of the length of the interval $[-M, M]$, and the "$1/16^q$" was designed to be safely smaller than the natural factor of $1/8^q$ which results from the partition process.

In the two dimensional case, we replace the "$1/2M$" by "$1/4M^2$" (area of a square), and replace the "$1/16^q$" by "$1/256^q$" (again safely smaller than the natural factor of $1/64^q$ which results from the partition process).

The trivial modifications for the case $q = 0$ are left to the reader.

(The algorithm in Section 4). Once the CompNorms have been found—cf. above—there is essentially no change in Section 4, other than the obvious fact that the single index i is replaced by a pair of indices i, j.

(Section 5, "Not an eigenvalue!") We examine the crucial lemma in step 4. in Section 5. In the proof of this lemma we had the statement: "There is some index i (which we label u) such that $\tau_{qu}(\lambda) \geq 1/2$." Now, because of the two-dimensional picture, the corresponding statement becomes: There is some pair of indices u, v such that $\tau_{quv}(\lambda) \geq 1/4$. In the previous proof, we went from $1/2$ down to $1/7$; the exact size of these constants did not matter: only the fact that the second was strictly less than the first. Here we can go from $1/4$ to $1/7$, so the constant $1/7$ still suffices in the complex case. [Of course, the point of the constant $1/7$ is that it is strictly greater than 1/8—since the "Not an eigenvalue!" declaration uses the cutoff value of 1/8.]

This completes our listing of the modifications. □

7. Unbounded Self-Adjoint Operators

Let $T: H \to H$ be an unbounded self-adjoint operator. It is well known (cf. Section 1) that $N = (T - i)^{-1}$ exists and is a bounded normal operator. Thus, in the classical (non-computable) case, the theory of unbounded self-adjoint operators reduces to that of bounded normal operators. A computable treatment of the spectrum for bounded normal operators was outlined in the preceding section. In this section, we extend this treatment to unbounded self-adjoint operators.

We recall the difinition. Let X be a Banach space with a computability structure. A closed operator T is *effectively determined* if there is a computable sequence of pairs $\{(e_n, Te_n)\}$ which spans a dense subspace of the graph of T. Then we also say that $\{e_n\}$ is an *effective generating set for T*.

The following trivial observation will be used in what follows. If T is effectively determined, if $\{e_n\}$ is an effective generating set for T, and if α is a computable real or complex constant, then the operator $T + \alpha$ is effectively determined, and $\{e_n\}$ is an effective generating set for $T + \alpha$. The proof is clear.

The following proposition, which we will use in our proof, may also be of independent interest.

Proposition. *Let $T: X \to X$ be effectively determined, and let $\{e_n\}$ be an effective generating set for T. Suppose that T^{-1} exists and is a bounded operator. Then T^{-1} is effectively determined.*

Proof. Since T^{-1} is bounded, it suffices to compute $T^{-1}e_n$ for any effective generating set $\{e_n\}$. More generally, we show how to compute $\{T^{-1}x_n\}$ for any computable sequence $\{x_n\}$. To do this we show how to compute $T^{-1}x$ for any computable x; it will be obvious that our procedure extends effectively to computable sequences $\{x_n\}$.

Since T^{-1} is a bounded operator, the range of T is the whole Banach space X. That is, the projection of the graph $\{(u, v): v = Tu\}$ onto the v-coordinate is X. Let $v = x$, so that $u = T^{-1}x$. Now to compute u effectively, we proceed as follows.

Let M be an integer such that $\|T^{-1}\| < M$. Let $\{p_i\}$ be an effective listing of all (real/complex) rational linear combinations of the e_n. Since T is effectively determined, the set of pairs (p_i, Tp_i) is dense in the graph of T. Hence for any k there exists an i such that

$$\|p_i - u\| + \|Tp_i - v\| < 2^{-k}/M,$$

which immediately gives

$$\|Tp_i - v\| < 2^{-k}/M.$$

To compute $u = T^{-1}x$ to within an error $<2^{-k}$, we set $v = x$, and then wait until a p_i turns up such that $\|Tp_i - x\| < 2^{-k}/M$. Then since $\|T^{-1}\| < M$, $\|p_i - T^{-1}x\| < 2^{-k}$. The u-coordinate p_i is our desired approximation to $T^{-1}x$. □

Now we return to the case of unbounded self-adjoint operators on a Hilbert space H. For convenience, we recall the result which we need to show.

Second Main Theorem, unbounded case, parts (i) and (ii). *Let $T: H \to H$ be an effectively determined unbounded self-adjoint operator. Then there exists a computable sequence $\{\lambda_n\}$ of real numbers, and a recursively enumerable set A of integers such that:*

each $\lambda_n \in \text{spectrum}(T)$;
the spectrum of T is the closure in $\mathbb{R}$ of the set $\{\lambda_n\}$;
the set of eigenvalues of T coincides with $\{\lambda_n: n \in \mathbb{N} - A\}$.

[Thus the results of Sections 1–5 extend to the case of unbounded self-adjoint operators.]

Proof. We combine the proposition above with Theorem 1, Section 6 (normal operators). Since T is self-adjoint, the operator $N = (T - i)^{-1}$ is bounded and normal. From the proposition above, we see that N is effectively determined. Now we apply Theorem 1 for normal operators. This asserts that there is a computable sequence of complex numbers $\{\mu_n\}$ which is dense in spectrum(N) and such that the eigenvalues of N consist of $\{\mu_n: n \in \mathbb{N} - A\}$ for some recursively enumerable set A.

Now the transition from N to T requires three steps.

1. Since $N = (T - i)^{-1}$, $T = N^{-1}(1 + iN)$. The spectrum/eigenvalues of N are mapped onto those of T by the function $\lambda = (1 + i\mu)/\mu$.

2. We observe that 0 is not an eigenvalue of N (else $T - i$ would map zero onto a nonzero vector). However, since T is unbounded, $0 \in \text{spectrum}(N)$, and 0 is a *limit point* of the spectrum of N.

3. We have to deal with the possibility that some of the $\mu_n = 0$. We simply delete these μ_n. The set B of n for which $\mu_n \neq 0$ is recursively enumerable (although perhaps not recursive). Consequently we keep $\{\mu_n: n \in B\}$, and replace the set A by $A \cap B$. Since 0 is a limit point of the spectrum, the set $\{\mu_n: \mu_n \neq 0\}$ is still dense in spectrum(N).

Combining steps 1. to 3. above, we see that the sequence $\{\lambda_n = (1 + i\mu_n)/\mu_n: n \in B\}$ fulfills the conditions of the theorem. □

8. Converses

Here we prove the converse parts (iii) and (iv) of the Second Main Theorem. We recall that (iii) is the converse to part (i), and (iv) is the converse to part (ii): the proof of the "positive" parts (i) and (ii) has occupied most of this chapter—Sections 1 to 7 above. Here, for convenience, we restate all of the parts (i) to (iv).

8. Converses

Consider an effectively determined (bounded or unbounded) self-adjoint operator $T: H \to H$. Part (i) asserts that there exists a computable real sequence $\{\lambda_n\}$ such that the spectrum of T coincides with the closure of $\{\lambda_n\}$. The converse part (iii) (Example 1 below) asserts that, given a computable real sequence $\{\lambda_n\}$, we can construct an operator T as above such that closure of $\{\lambda_n\}$ coincides with the spectrum of T. We recall that, since the spectral norm of T equals its norm, if the set $\{\lambda_n\}$ is bounded then T is bounded.

Now again consider an effectively determined (bounded or unbounded) self-adjoint operator $T: H \to H$. Part (ii) asserts that there exists a computable real sequence $\{\lambda_n\}$ and a recursively enumerable set A of natural numbers, such that the set of eigenvalues of T coincides with $\{\lambda_n : n \notin A\}$. The converse part (iv) (Example 2 below) asserts that given $\{\lambda_n\}$ and A as above, we can construct an effectively determined self-adjoint operator T such that the set $\{\lambda_n : n \notin A\}$ coincides with the set of eigenvalues of T. When the set $\{\lambda_n\}$ is bounded, the operator T can be chosen to be bounded.

Example 1. Let $\{\lambda_n\}$ be a computable sequence of real numbers. There exists an effectively determined, self-adjoint operator T whose spectrum is the closure of $\{\lambda_n\}$ in $\mathbb{R}$.

Proof. Let $\{e_n\}$ be a computable orthonormal basis for the Hilbert space H (cf. Lemma 7, Section 6 in Chapter 4). In terms of this basis, T is the self-adjoint operator defined by the matrix:

$$T \sim \begin{pmatrix} \lambda_0 & & & & & \\ & \lambda_1 & & & & \\ & & \lambda_2 & & 0 & \\ & & & \ddots & & \\ & & 0 & & \lambda_n & \\ & & & & & \ddots \end{pmatrix}.$$

This means that $Te_n = \lambda_n e_n$, and the domain of T is the set of all vectors $x = \sum c_n e_n$ for which both of the sums $\|x\|^2 = \sum |c_n|^2$ and $\|Tx\|^2 = \sum |\lambda_n c_n|^2$ are finite. It is easy to verify (cf. Riesz and Sz-Nagy [1955]) that T, as so defined, is self-adjoint. Since $\{\lambda_n\}$ is computable, the sequence of pairs $\{\langle e_n, Te_n \rangle\} = \{\langle e_n, \lambda_n e_n \rangle\}$ is computable and forms an effective generating set for the graph of T. Hence T is effectively determined.

The eigenvalues $\lambda_n \in \text{spectrum}(T)$, and since $\text{spectrum}(T)$ is closed, the closure of $\{\lambda_n\}$ is a subset of $\text{spectrum}(T)$. To show that $\text{closure}\{\lambda_n\} = \text{spectrum}(T)$, consider any real number $\alpha \notin \text{closure}\{\lambda_n\}$. Then the sequence of numbers $\{1/(\lambda_n - \alpha)\}$ is bounded. These numbers form the elements of the diagonal matrix for $(T - \alpha I)^{-1}$. Thus $(T - \alpha I)^{-1}$ exists and is bounded. Hence $\alpha \notin \text{spectrum}(T)$. □

Example 2. Let $\{\lambda_n\}$ be a computable sequence of real numbers. Let A be a recursively enumerable set of natural numbers. Then there exists an effectively determined self-adjoint operator T such that the set of eigenvalues of T coincides with the set $\{\lambda_n: n \notin A\}$. When the sequence $\{\lambda_n\}$ is bounded, the operator T is also bounded.

Here we give the construction only for the case where $\{\lambda_n\}$ is bounded—so that the resulting operator T is also bounded. The extension to unbounded $\{\lambda_n\}$ is purely mechanical. In fact, the construction for unbounded $\{\lambda_n\}$ is identical to that given here. However, the verifications (e.g. that T is self-adjoint) require a number of technicalities associated with unbounded self-adjoint operators. These facts about unbounded operators can be found e.g. in Riesz and Nagy [1955], p. 314. We add that, when $\{\lambda_n\}$ is bounded, these technical difficulties disappear. (We also mention in passing that the example in part (iii) above was done for both the bounded and unbounded case.) Now here is the construction.

Proof for bounded $\{\lambda_n\}$. This construction is an extension of the preceding one, but is a little more complicated. We let H be a countable direct sum of spaces H_n isomorphic to $L^2[-1, 1]$. Let e_{nm} be the function on the n-th copy of $L^2[-1, 1]$ given by $e_{nm}(x) = (1/\sqrt{2})e^{\pi imx}$, $m = 0, \pm 1, \pm 2, \ldots$. Then $\{e_{nm}\}$ is a computable orthonormal basis for H.

As a preliminary step, we begin with the operator T_0 defined by

$$T_0 f = \lambda_n f \quad \text{for } f \in H_n.$$

Thus, T_0 restricted to the n-th copy of $L^2[-1, 1]$ coincides with multiplication by the constant λ_n. This makes λ_n an eigenvalue of T_0.

The idea behind our construction is that, by perturbing $T_0 | H_n$ ever so slightly, we can destroy the eigenvalue λ_n and replace it by a narrow band of continuous spectrum. This perturbation can come at any stage; the later it comes, the smaller it will be.

We now give the details. Let $a(k)$ be a 1-1 recursive function which enumerates the set A. We start with the operator T_0 defined above. At the k-th stage, we introduce the following perturbation. Let T_k denote the operator as it is before the k-th stage. Then we define T_{k+1} by:

$$T_{k+1} = T_k \text{ on the orthocomplement of } H_{a(k)} \text{ in } H.$$

On $H_{a(k)}$, we change

$$T_k = \text{multiplication by the constant } \lambda_{a(k)}$$

into

$$T_{k+1} = \text{multiplication by the } \textit{function } \lambda_{a(k)} + 2^{-k}x.$$

The resulting operator T_{k+1} on $H_{a(k)}$ has the form:

$$T_{k+1}[f(x)] = (a + bx) \cdot f(x),$$

8. Converses

where $a = \lambda_{a(k)}$, $b = 2^{-k} \neq 0$. It is well known and easy to verify that such an operator has only continuous spectrum, and no eigenvalues.

(The spectrum of T_{k+1} on $H_{a(k)}$ coincides with the range of $(a + bx)$ on $[-1, 1]$, i.e. with the interval between $\lambda_{a(k)} \pm 2^{-k}$. Thus the eigenvalue $\lambda_{a(k)}$ is replaced by a band of continuous spectrum with a band width of $2 \cdot 2^{-k}$.)

Let $T = \lim T_k$. We must verify that the operator T is effectively determined. Since T is bounded, it suffices to show that $\{T(e_{nm})\}$ is computable, where $\{e_{nm}\}$ is the effective generating set given above. Now it is clear that the triple sequence $\{T_k(e_{nm})\}$ is computable in k, n, m. Since $\|T_k - T_{k-1}\| \leq 2^{-k}$, the operators T_k converge uniformly and effectively to T as $k \to \infty$. Hence by the Limit Axiom $\{T(e_{nm})\}$ is computable. Thus T is effectively determined.

Finally, the eigenvalues of T are precisely the set of λ_n not destroyed, i.e. $\{\lambda_n : n \notin A\}$. This completes the example and finishes the proof of the Second Main Theorem. □

Addendum: Open Problems

Open problems abound. Here we merely give a representative sample. The problems fall into two categories—logic and analysis.

We begin with questions motivated by logic.

1. Our first problem is concerned with the following: What is the degree of difficulty of those analytic processes which have been proved to the computable? One topic of broad scope is to bring into analysis the whole complex of problems associated with $P = NP$ (cf. Cook [1971], Karp [1972], Friedman, Ko [1982], Friedman [1984], Ko [1983], Blum, Shub, Smale [to appear]). Thus we may ask which analytic processes are computable in polynomial time, polynomial space, exponential time, etc. In the same manner, we can ask about levels of difficulty within the Grzegorczyk hierarchy, or any other subrecursive hierarchy. Or we could fix our attention on the primitive recursive functions. There is no reason to believe that the answers to these questions will be automatic extensions of the general recursive case.
2. For processes proved to be noncomputable, we can also ask for fine structure—this time via the theory of degrees of unsolvability. Most of the noncomputability results in this book make use of an arbitrary recursively enumerable nonrecursive set. In fact, any recursively enumerable nonrecursive set—of any degree of unsolvability—will do. The question is: Can we replace results which merely assert that a certain process is noncomputable by a fine structure for that process, involving different degrees of unsolvability?
3. Our third problem is concerned with nonclassical reasoning. We recall that the reasoning in this book is classical—i.e. the reasoning used in everyday mathematical research. This contrasts with the intuitionist approach (e.g. of Brouwer), the constructivist approach (e.g. of Bishop), and the Russian school (e.g. Markov and Šanin). A natural question is: What are the analogs, within these various modes of reasoning, of the results in this book?

    In this connection, we cite the work of Feferman [1984], who originated the system T_0 for representing Bishop-style constructive mathematics. T_0 has both constructive and classical models. In particular, Feferman reformulated our First Main Theorem in T_0, and left as an open question the status of our Second Main Theorem.
4. Our fourth problem concerns higher order recursion theory. Let us set the stage.

    Higher order recursion theory, of course, deals with functionals of functions from $\mathbb{N}$ to $\mathbb{N}$, functionals of such functionals, etc. A functional approach to recursive analysis was given by Grzegorczyk [1955]. Here the real numbers are viewed as the set $\mathcal{R}$ of functions $\varphi\colon \mathbb{N} \to \mathbb{N}$. A function of a real variable is associated with a functional Φ mapping $\mathcal{R}$ into $\mathcal{R}$. The function of a real variable is "computable" if the associated functional Φ is computable—i.e. a general recursive functional.

Grzegorczyk proved that the functional approach is equivalent to the notion of Chapter-0-computability for continuous functions (Grzegorczyk [1957]). Chapter-0-computability, since it is tied more closely to standard analytic concepts (e.g. effective uniform continuity), appears to be more amenable to work in analysis.

The Chapter-0 definition leads in a natural way to certain generalizations—e.g. the definition of L^p-computability, and beyond that, the concept of a computability structure. The main results in this book are based on these generalizations. Higher order recursion theory leads to its own generalizations. The problem is: How do the concepts and results of higher order recursion theory relate to the concepts and results developed in this book?

We turn now to analysis. Here the set of problems is almost limitless, and we give only a few samples, which we find particularly appealing.

5. This problem combines ideas from topology, complex analysis, and recursion theory. Recursive topology provides a standard definition of "computable" open set in $\mathbb{R}^n$. However, it will emerge that this traditional definition may not be the right one.

Let us consider this issue from the viewpoint of complex analysis. For this purpose, of course, we work within the 2-dimensional complex plane. We begin with a simply connected proper open subset of the plane. Here the crucial theorem is the Riemann Mapping Theorem, which asserts that every such region has a conformal mapping onto the open unit disk. This mapping is, up to trivial transformations, unique. Two obvious questions are: (1) If the mapping is computable, is the region computable? (2) If the region is computable, is the mapping computable? This in turn leads to the question of what we should mean by a computable region in the plane.

The standard definition of a "computably open set" Ω goes as follows: Ω is *computably open* if it is the union of a sequence of disks $\{|z - a_i| < r_i\}$, where $\{r_i\}$ and $\{a_i\}$ are respectively computable sequences of real and complex numbers (Lacombe [1957b], Lacombe/Kreisel [1957]). This definition will not suffice for complex analysis. For, if we adopted it, the hope for any connection between computable region and computable function would be dashed. Here is a trivial counterexample. Consider the region $\Omega = \{z\colon |z| < \alpha\}$, where α is a noncomputable real which is the limit of a computable monotone sequence of rationals. Then the region Ω would be computably open in the sense defined above. But the natural conformal mapping onto the unit disk, namely z/α, is obviously not computable.

The resolution of this question might provide an interesting interplay between plane set topology, complex analysis, and logic. The topological aspects could obviously be generalized to $\mathbb{R}^n$. For the analytic aspects, we might consider conformal mappings of multiply connected regions. Finally, one could generalize these problems to Riemann surfaces.

The above discussion applies, of course, to recursive analysis with the usual classical reasoning. We mention that a discussion within the constructivist framework appears in Bishop [1967] and Bishop/Bridges [1985].

6. Many of the theorems in this book deal with the computability aspects of linear analysis. There are still many unsolved problems in this area. Here we mention two.

Our first question is open ended. Throughout this book we have attempted to give general principles from which the effectivization or noneffectiveness of well-known classical theorems follows as corollaries. Obviously this program can be broadened in many ways. For example, it would be interesting to have a general principle which gave as a corollary the known facts concerning the Hahn-Banach Theorem. The facts are these. Metakides, Nerode, and Shore [1985] have proved a recursive version of the Hahn-Banach Theorem, in which they enlarge the norm of the functional by an arbitrary ε. They show that this enlargement is necessary. One particular question, which we would like to see emerge as an outgrowth of a more general principle, is the following. Characterize those Banach spaces for which we can obtain a recursive Hahn-Banach Theorem without an enlargement of the norm.

Our second question is: Under what conditions are the eigenvalues of a bounded effectively determined operator T computable? For compact operators an affirmative answer is well-known. When T is self-adjoint or normal an affirmative answer is provided by our Second Main Theorem. On the other hand when T is neither self-adjoint nor compact, noncomputable eigenvalues can occur (Theorem 4.5). However, for many nonnormal operators, the eigenvalues are known to be computable—indeed they have been computed. The problem, then, is to find conditions, more general than normality, which cover important applications and imply that the eigenvalues are computable.

7. Finally, we give some open problems concerned with nonlinear analysis. So far as we know, the only major nonlinear problem which has been investigated from the viewpoint of recursion theory is the Cauchy-Peano existence theorem for ordinary differential equations (Aberth [1971], Pour-El/Richards [1979]). Nonlinear analysis is a vast area, and its connections with recursion theory, at the time of this writing, remain largely untouched.

In many nonlinear problems, when they are dealt with classically, the technique of linearization plays an important role. This leads then to two questions. The first, absolutely untouched so far as we know, is the connection between the computability of the original nonlinear operator and the linear operator which results from it. The second concerns the computability of the eigenvalues of these linear operators. For self adjoint operators, this question has been answered by our Second Main Theorem. But for operators which are neither self-adjoint nor compact, the question remains open (cf. problem 6, above).

Another problem is to extend the First Main Theorem to nonlinear operators. More precisely, we might ask to what extent, and under what side conditions, the First Main Theorem holds?

A third area is the recursion theoretic study of particular nonlinear problems of classical importance. Examples are the Navier-Stokes equation, the KdV equation, and the complex of problems associated with Feigenbaum's constant.

Obviously, this discussion provides but a small sample of the questions which can be asked in recursive nonlinear analysis.

Bibliography

This bibliography contains, besides references cited in the book, representative samples of other work in the area. Although a few sources in the intuitionist or constructivist traditions of Brouwer, Bishop, the Russian school etc. are listed, we have mainly limited ourselves to works in which the usual classical reasoning of mathematics is employed.

Aberth, O.
1971   *The failure in computable analysis of a classical existence theorem for differential equations*, Proc. Amer. Math. Soc. 30, 151–156
1980   *Computable analysis*, McGraw-Hill: New York

Ahlfors, L.V.
1953   *Complex Analysis*, McGraw-Hill: New York

Baez, J.C.
1983   *Recursivity in quantum mechanics*, Trans. Amer. Math. Soc. 280, 339–350

Beeson, M.J.
1985   *Foundations of constructive mathematics*, Springer: Heidelberg, New York

Bishop, E.A.
1967   *Foundations of constructive analysis*, McGraw-Hill: New York
1970   *Mathematics as a numerical language,* Intuitionism and Proof Th.; 1968 Buffalo 53–71; ed. by Kino, A., Myhill, J., Vesley, R.E.. North Holland: Amsterdam

Bishop, E.A., Bridges, D.S.
1985   *Constructive analysis*, Springer: Heidelberg, New York

Blum, L., Shub, M., Smale, S.
     On a theory of computation and complexity over the real numbers: NP completeness, recursive functions and universal machines (to appear in Bull. Amer. Math. Soc.)

Bridges, D.S.
1979   *Constructive functional analysis*, Pitman Publ.: Belmont, London v + 203pp

Brouwer, L.E.J.
1908   *The unreliability of the logical principles (Dutch)*, Tijdsch. Wijsbegeerte 2,152–158; transl. in: Coll. Works I, 107–111 (1975) (English)
1975   *Collected works I*, North Holland: Amsterdam
1981   *Brouwer's Cambridge lectures on intuitionism*, Cambridge Univ. Pr.: Cambridge, GB

Caldwell, J., Pour-El, M.B.
1975   *On a simple definition of computable function of a real variable—with applications to functions of a complex variable*, Z. Math. Logik Grundlagen Math. 21, 1–19

Cleave, J.P.
1969   *The primitive recursive analysis of ordinary differential equations and the complexity of their solutions*, J. Comp. Syst. Sci. 3, 447–455

Cook, S.A.
1971   *The complexity of theorem proving procedures*, ACM Symp. Th. of Comput. (3); 1971 Shaker Heights 151–158

Cutland, N.J.
1980   *Computability. An introduction to recursive function theory*, Cambridge Univ. Pr.: Cambridge, GB

Dalen van, D.
1973   *Lectures on intuitionism*, Cambridge Summer School Math. Log.; 1971 Cambridge, GB 1–94; ed. by Rogers, H., Mathias, A.R.D.. Lect. Notes Math. 337, Springer: Heidelberg, New York

Davis, M.D.
1958   *Computability and unsolvability*, McGraw-Hill: New York

Denef, J., Lipshitz, L.
1984   *Power series solutions of algebraic differential equations*, Math. Ann. 267, 213–238

Dieudonné, J.A.
1969   *Foundations of modern analysis*, Academic Pr.: New York

Dunford, N., Schwartz, J.T.
1958   *Linear Operators, Part I*, Intersci. Publ.: New York

Feferman, S.
1977   *Theories of finite type related to mathematical practice*, Handb. of Math. Logic 913–971; ed. by Barwise, J.. Stud. Logic Found. Math. 90, North Holland: Amsterdam
1979   *Constructive theories of functions and classes*, Logic Colloq.; 1978 Mons 159–224; ed. by Boffa, M., Dalen van, D., McAloon, K.. Stud. Logic Found. Math. 97, North Holland: Amsterdam
1984   *Between constructive and classical mathematics*, Computation and Proof Theory, Logic Colloq.; 1983 Aachen 2, 143–162; ed. by Börger, E., Oberschelp, W., Richter, M.M., Schinzel, B., Thomas, W.. Lect. Notes Math. 1104, Springer: Heidelberg, New York

Friedman, H.M., Ko, Ker-I
1982   *Computational complexity of real functions*, Theor. Comput. Sci. 20, 323–352

Friedman, H.M.
1984   *The computational complexity of maximization and integration*, Adv. Math. 53, 80–98

Goodstein, R.L.
1961   *Recursive analysis*, North Holland: Amsterdam

Grzegorczyk, A.
1955   *Computable functionals*, Fund. Math. 42, 168–202
1957   *On the definitions of computable real continuous functions*, Fund. Math. 44, 61–71
1959   *Some approaches to constructive analysis*, Constructivity in Math.; 1957 Amsterdam 43–61; ed. by Heyting, A.. Stud. Logic Found. Math., North Holland: Amsterdam

Halmos, P.R.
1950   *Measure theory*, Van Nostrand: New York
1951   *Introduction to Hilbert space*, Chelsea: New York

Hauck, J.
1981   *Berechenbarkeit in topologischen Räumen mit rekursiver Basis*, Z. Math. Logik Grundlagen Math. 27, 473–480
1984   *Zur Wellengleichung mit konstruktiven Randbedingungen*, Z. Math. Logik Grundlagen Math. 30, 561–566

Hellwig, G.
1964   *Partial differential equations*, Blaisdell: New York

Heyting, A.
1956   *Intuitionism. An introduction*, North Holland: Amsterdam

Huang, W., Nerode, A.
1985   *Application of pure recursion theory in recursive analysis (Chinese)*, Shuxue Xuebao 28/5, 625–635

Hyland, J.M.E.
1982   *Applications of constructivity*, Int. Congr. Log., Meth. and Phil. of Sci. (6, Proc.); 1979 Hannover 145–152; ed. by Cohen, L.J., Łos, J., Pfeiffer, H., Podewski, K.-P.. Stud. Logic Found. Math. 104, North Holland: Amsterdam

Kalantari, I., Weitkamp, G.
1985   Effective topological spaces I: A definability theory. II: A hierarchy, Ann. Pure Appl. Logic 29, 1–27, 207–224

Karp, R.M.
1972   Reducibility among combinatorial problems, Compl. of Computer Computation; 1972 Yorktown Heights 85–103; edited by Miller, R.E., Thatcher, J.W., Bohlinger, J.D.. Plenum Publ.: New York

Klaua, D.
1961   Konstruktive Analysis, Dt. Verlag Wiss.: Berlin

Kleene, S.C.
1952   Introduction to metamathematics, North Holland: Amsterdam
1956   A note on computable functions, Indag. Math. 18, 275–280

Ko, Ker-I
1983   On the computational complexity of ordinary differential equations, Inform. and Control 58, 157–194

Kreisel, G., Lacombe, D.
1957   Ensembles récursivement mesurables et ensembles récursivement ouverts ou fermés, C. R. Acad. Sci., Paris 245, 1106–1109

Kreisel, G.
1958   Review of "Meschkowski, H.: 'Zur rekursiven Funktionentheorie', Acta Math. 95(1956), 9–23", MR 19, 238
1974   A notion of mechanistic theory, Synthese 29, 11–16
1982   Brouwer's Cambridge lectures on intuitionism, Canad. Phil. Rev. 2, 249–251

Kreitz, C., Weihrauch, K.
1984   A unified approach to constructive and recursive analysis, Computation and Proof Theory, Logic Colloq.; 1983 Aachen 2, 259–278; edited by Boerger, E., Oberschelp, W., Richter, M.M., Schinzel, B. Thomas, W., Lect. Notes Math. 1104, Springer: Heidelberg, New York

Kushner, B.A.
1973   Lectures on constructive mathematical analysis (Russian), Nauka: Moskva 447pp; transl.: Amer. Math. Soc.: Providence (1984) (English)

Lachlan, A.H.
1963   Recursive real numbers, J. Symb. Logic 28, 1–16

Lacombe, D.
1955a   Extension de la notion de fonction récursive aux fonctions d'une ou plusieurs variables réelles. I, C. R. Acad. Sci., Paris 240, 2478–2480
1955b   Extension de la notion de fonction récursive aux fonctions d'une ou plusieurs variables réelles II, III, C. R. Acad. Sci., Paris 241, 13–14, 151–153
1955c   Remarque sur les opérateurs récursifs et sur les fonctions récursives d'une variable réelle, C. R. Acad. Sci., Paris 241, 1250–1252
1957a   Quelques propriétés d'analyse récursive, C. R. Acad. Sci., Paris 244, 838–840, 996–997
1957b   Les ensembles récursivement ouverts ou fermés, et leurs applications à l'analyse récursive, C. R. Acad. Sci., Paris 245, 1040–1043
1958   Sur les possibilités d'extension de la notion de fonction récursive aux fonctions d'une ou plusieurs variables réelles, Raisonn. en Math. et Sci. Expér.; 1955 Paris 67–71, Colloq. Int. CNRS. 70, CNRS Inst. B. Pascal: Paris 140pp
1959   Quelques procédés de définition en topologie récursive, Constructivity in Math.; 1957 Amsterdam 129–158 ed. by Heyting, A.. Stud. Logic Found. Math., North Holland: Amsterdam

Loomis, L.H.
1953   An introduction to abstract harmonic analysis, Van Nostrand: New York

Mazur, S.
1963   Computable analysis, PWN: Warsaw

Metakides, G., Nerode, A.
1982   The introduction of non-recursive methods into mathematics, Brouwer Centenary Symp.; 1981 Noordwijkerhout 319–335; in Troelstra, A.S., Dalen van, D. (eds.) (1982)

Metakides, G., Nerode, A., Shore, R.A.
1985   *Recursive limits on the Hahn-Banach theorem*, E. Bishop—Reflection on Him & Research; 1983 San Diego 85–91; edited by Rosenblatt, M.. Contemp. Math. 39, Amer. Math. Soc.: Providence

Moschovakis, Y.N.
1964   *Recursive metric spaces*, Fund. Math. 55, 215–238
1966   *Notation systems and recursive ordered fields*, Compos. Math. 17, 40–71

Mostowski, A.
1957   *On computable sequences*, Fund. Math. 44, 37–51

Mycielski, J.
1981   *Analysis without actual infinity*, J. Symb. Logic 46, 625–633

Myhill, J.R.
1953   *Criteria of constructibility for real numbers*, J. Symb. Logic 18, 7–10
1971   *A recursive function, defined on a compact interval and having a continuous derivative that is not recursive*, Michigan Math. J. 18, 97–98

Orevkov, V.P.
1963   *A constructive mapping of the square onto itself displacing every constructive point (Russian)*, Dokl. Akad. Nauk SSSR 152, 55–58; transl. in: Sov. Math., Dokl. 4, 1253–1256 (1963) (English)

Petrovskij, I.G.
1967   *Partial differential equations*, Saunders: Philadelphia

Pour-El, M.B.
1974   *Abstract computability and its relation to the general purpose analog computer (some connections between logic, differential equations and analog computers)*, Trans. Amer. Math. Soc. 199, 1–28

Pour-El, M.B., Richards, I.
1979   *A computable ordinary differential equation which possesses no computable solution*, Ann. Math. Logic 17, 61–90
1981   *The wave equation with computable initial data such that its unique solution is not computable*, Adv. Math. 39, 215–239
1983a   *Computability and noncomputability in classical analysis*, Trans. Amer. Math. Soc. 275, 539–560
1983b   *Noncomputability in analysis and physics: a complete determination of the class of noncomputable linear operators*, Adv. Math. 48, 44–74
1984 L^p-*computability in recursive analysis*, Proc. Amer. Math. Soc. 92, 93–97
1987   *The eigenvalues of an effectively determined self-adjoint operator are computable, but the sequence of eigenvalues is not*, Adv. Math. 63, 1–41
        Computability on a Banach space—the eigenvector theorem, (to appear)

Remmel, J.B.
1978   *Recursively enumerable boolean algebras*, Ann. Math. Logic 15, 75–107

Rice, H.G.
1954   *Recursive real numbers*, Proc. Amer. Math. Soc. 5, 784–791

Richardson, D.B.
1968   *Some undecidable problems involving elementary functions of a real variable*, J. Symb. Logic 33, 514–520

Riesz, F., Sz.-Nagy, B.
1955   *Functional analysis*, Ungar: New York and London

Robinson, R.M.
1951   *Review of "Peter, R.: 'Rekursive Funktionen', Akad. Kiado, Budapest, 1951"*, J. Symb. Logic 16, 280

Rogers Jr., H.
1967   *Theory of recursive functions and effective computability*, McGraw-Hill: New York

Rosenbloom, P.C.
1945 *An elementary constructive proof of the fundamental theorem of algebra*, Amer. Math. Mon. 52, 562–570

Rudin, W.
1973 *Real and complex analysis*, McGraw-Hill: New York

Ščedrov, A.
1984 *Differential equations in constructive analysis and in the recursive realizability topos*, J. Pure Appl. Algebra 33, 69–80

Schütte, K.
1977 *Proof theory*, Springer: Heidelberg, New York

Shanin, N.A.
1962 *Constructive real numbers and constructive function spaces (Russian)*, Tr. Mat. Inst. Steklov 67, 15–294; Akad. Nauk SSSR: Moskva; transl.: Amer. Math. Soc.: Providence (1968) (English)

Shepherdson, J.C.
1976 *On the definition of computable function of a real variable*, Z. Math. Logik Grundlagen Math. 22, 391–402

Shohat, J.A., Tamarkin, J.D.
1943 *The problem of moments*, Amer. Math. Soc.: Providence

Simpson, S.G.
1984 *Which set existence axioms are needed to prove the Cauchy/Peano theorem for ordinary differential equations?*, J. Symb. Logic 49, 783–802

Soare, R.I.
1969 *Recursion theory and Dedekind cuts*, Trans. Amer. Math. Soc. 140, 271–294
1987 *Recursively enumerable sets and degrees*, Springer: Heidelberg, New York

Specker, E.
1949 *Nicht konstruktiv beweisbare Sätze der Analysis*, J. Symb. Logic 14, 145–158
1959 *Der Satz vom Maximum in der rekursiven Analysis*, Constructivity in Math.; 1957 Amsterdam 254–265; edited by Heyting, A. Stud. Logic Found. Math., North Holland: Amsterdam
1969 *The fundamental theorem of algebra in recursive analysis*, Constr. Aspects Fund. Thm. Algeb.; 1967 Zürich 321–329; edited by Dejon, B., Henrici, P.. Wiley and Sons: New York

Traub, J.F., Wasilkowski, G.W., Woźniakowski, H.
1988 *Information-based complexity*, Academic Press Inc., Boston, New York

Troelstra, A.S.
1969 *Principles of intuitionism*, Springer: Heidelberg, New York

Troelstra, A.S., Dalen van D. (eds.)
1982 *The L.E.J. Brouwer centenary symposium*, Stud. Logic Found. Math. 110, North Holland: Amsterdam

Tsejtin, G.S.
1959 *Algorithmic operators in constructive complete separable metric spaces (Russian)*, Dokl. Akad. Nauk SSSR 128, 49–52
1962 *Mean value theorems in constructive analysis (Russian)*, Tr. Mat. Inst. Steklov 67, 362–384; Akad. Nauk SSSR: Moskva; transl. in: Amer. Math. Soc., Transl., Ser. 2 98, 11–40 (1971) (English)

Tsejtin, G.S., Zaslavskij, I.D.
1962 *On singular coverings and properties of constructive functions connected with them (Russian)*, Tr. Mat. Inst. Steklov 67, 458–502; Akad. Nauk SSSR: Moskva; transl. in: Amer. Math. Soc., Transl., Ser. 2 98, 41–89 (1971) (English)

Turing, A.M.
1936 *On computable numbers, with an application to the "Entscheidungsproblem"*, Proc. London Math. Soc., Ser. 2 42, 230–265; corr. ibid. 43, 544–546 (1937)

Weyl, H.
1918 *Das Kontinuum*, Veit: Leipzig

Wolfram, S.
1985 *Undecidability and intractability in theoretical physics*, Phys. Rev. Lett. 54, 735–738

Zaslavskij, I.D.
1955 *Disproof of some theorems of classical analysis in constructive analysis (Russian)*, Usp. Mat. Nauk 10/4, 209–210
1962 *Some properties of constructive real numbers and constructive functions (Russian)*, Tr. Mat. Inst. Steklov 67, 385–457 Akad. Nauk SSSR: Moskva; transl. in: Amer. Math. Soc., Transl., Ser. 2 57, 1–84 (1966) (English)

Zygmund, A.
1959 *Trigonometric series. Vol. I, II*, Cambridge Univ. Pr.: Cambridge, GB

Subject Index

A
Aberth, 5, 194
ad hoc computability [*see* computability, counterexamples]
adjoint (definition), 125–126
Ahlfors, 61
analytic continuation, 60–64
analytic functions, 50–51, 59–64 [*see also* computability]
axioms (for computability)
— Banach space, 1, 3, 5, 11, 77–82 [*Note.* The axioms are used in the proof of virtually every theorem from Chapter 2 onwards. This is understood. We do not list all the pages, since such a list would grow so long that it would cease to be informative.]
— Banach space (statement), 81
— computability structure, 1, 3, 5, 77–82, 85–87 [*Note.* Again, this notion permeates the book. Only the most important references are listed.]
— computability structure (definition), 80
— Hilbert space [same as for a Banach space]

B
Banach/Mazur [*see* computability]
Banach space [*see* axioms, computability, First Main Theorem]
Banach space (definition), 8
Bishop, 4, 192, 194
Blum, 192
bounded operators, 1–2, 93–94, 96, 123, 128, 150–184 [*see also* First Main Theorem]
bounded operators (definition), 96
Bridges, 194
Brouwer, 4, 192

C
C, 8 [*see also* computability, Chapter 0]
C^n, 8 [*see also* computability]
C^∞, 8 [*see also* computability]
C_o, 8 [*see also* computability]
Caldwell, 12, 26, 62 [*see also* computability]
Cauchy integral formula, 12, 60
characteristic function (definition), 8
Closed Graph Theorem (classical, non-effective), 97, 108
closed operators, 93–94, 96–100 [*see also* First Main Theorem]
closed operators (definition), 96–97
Closure Criterion (First/Second), 98–100, 105, 108, 110, 116
CompNorm (in proof of the Second Main Theorem), 165–166
compact operators, 123, 129, 133, 136
comparisons (between real numbers/ rationals), 14–15, 23
Composition Property, 81
computability [For the underlying notion of computability on a Banach space, see "axioms, Banach space." The derived notions, for standard Banach spaces and related topics, are listed below. Theorems are listed elsewhere.]
— ad hoc, 5, 80, 90–92, 124, 134–142, 146–147
— analytic functions, 59–60
— Banach/Mazur, 28, 64–65
— Banach space [*see* axioms]

— C^n, 50–59, 104–105, 117
— C^∞, 50, 54–57, 60
— C_o, 84, 91, 107–108, 111, 118–119
— Caldwell/Pour-El, 12, 24–28 [see also Definition B and Chapter 0 computabity]
— Chapter 0 (an alternative designation for computability in the sense of Definitions A and B), 50, 79, 82–83, 94–95, 104–106, 115–116, 118–120, 149, 161–162
— complex numbers 14, 27
— continuous functions [same as Chapter 0 computability]
— Definition A, 12, 25–28, 28 (again in Section 4), 33, 36–37, 40, 44
— Definition B, 12, 25–28, 33, 36–37, 39, 44
— Definitions A and B (equivalence of), 36–37, 44–49
— double sequence, 18, 80
— energy norm, 95–96, 116–118
— Grzegorczyk/Lacombe, 4–5, 12, 24–28 [see also Definition A and Chapter 0 computability]
— Hilbert space [see axioms]
— inner products, 136–138
— intrinsic, 79, 82–85, 90–91
— L^p, 5, 83–85, 94–95, 107–114
— ℓ^p, 85, 94, 107–111
— L^∞, 89–90
— ℓ_o^∞, 85, 107–108, 110
— open sets, 193
— operators [see effectively determined operators]
— real numbers, 11, 13–17, 20
— rectangles, 25, 27
— sequences of rational numbers, 12, 14, 24
— sequences of real numbers, 11–12, 17–24
— sequences on a Banach space [see axioms]
— sequential (for continuous functions), 25, 28–29, 32, 34–35, 40, 51, 55, 64, 67–68, 71–72
— Sobolev spaces, 5, 95–96
— structure [see axioms]
— $(0, \infty)$, 27–28
continuous functions [see computability]
convolution, 69–70
counterexamples (involving)
— ad hoc computability, 90–92, 134–139, 146

— analytic continuation for noncompact domains/sequences of functions, 62–63
— converse parts (iii) and (iv) of the Second Main Theorem, 188–191
— creation and destruction of eigenvalues, 130–132
— derivatives, 51, 55, 58–59, 104–105
— derivatives of a sequence of functions, 59
— eigenvectors [see Eigenvector Theorem]
— entire functions, 62
— Fourier series, 105, 110
— Intermediate Value Theorem for sequences of functions, 42
— isometry of nonequivalent computability structures, 146
— L^∞, 89–90
— non closed operators, 105
— non normal operators, 132–133
— noncompact domains, 58, 62
— noncomputable real numbers, 12, 20, 129–130, 132–133, 135
— noneffective convergence, 11–12, 16, 19–20, 22–23, 105
— norm of an effectively determined operator, 129
— separable but not effectively separable Banach spaces, 88
— sequences of eigenvalues, 129, 189–191
— sequences of n-th derivatives, 55
— sequences of real numbers, 19–20, 22–24
— sequences of step functions, 112
— sequentially computable continuous functions, 67
— unbounded operators [see First Main Theorem]
— wave equation, 68–69, 72–73, 115–116, 120
Cook, 192
Cutland, 7

D

$\mathscr{D}(T)$ [see domain]
Davis, 7
degrees of unsolvability, 192
density (effective), 82, 85–87
density (not necessarily effective), 82, 85, 88
differentiation, 11, 40, 50–59, 60–62, 104–105

Subject Index

Dirichlet norm [see computability, energy norm]
distributions, 99–100, 108, 116
domain of an operator, 96–104, 125–126
Dunford/Schwartz, 110

E

effective
— convergence, 11–24, 34, 37, 44–49, 56, 72, 81, 86–87, 105–106, 162
— convergence (definitions), 14, 18, 34, 81
— generating set, 78–79, 82–87, 93–94, 101, 104, 127 [Note. This notion permeates the book. Only the most important references are listed.]
— generating set (definition), 82
— uniform continuity, 25–27, 29, 32, 34–35, 40, 50, 53–55, 64, 67–68, 71–72
— uniform continuity (definition), 25–27

effectively
— determined operator, 2, 123–124, 127–129, 132–134, 138–139, 150, 158–160, 184–185, 187–188
— determined operator (definition), 127
— separable, 78–79, 82, 88–89, 128, 138, 141–142
— separable (definition), 82

eigenvalue, 2, 123–124, 126–127 [see also Second Main Theorem]
eigenvalue (definition), 127
eigenvector, 2, 123–124, 126–127 [see also Eigenvector Theorem]
eigenvector (definition), 127
Eigenvector Theorem, 1–4, 77, 80, 123–124, 133–142
Eigenvector Theorem (statement), 133–134
elementary functions, 21
energy norm [see computability]
exponential time, 192
extremal points, 148

F

Feferman, 192
Feigenbaum's constant, 194
First Main Theorem 1–4, 69, 77, 93–96, 101–120
First Main Theorem (statement), 101
Fourier series, 83, 94–95, 105–106, 108–111
Fourier transforms, 3, 95, 108–111
Friedman, 192

G

Gödel, 78
Grzegorczyk, 4, 5, 12, 25, 28, 104, 192–193 [see also computability]
Grzegorczyk hierarchy, 192

H

Hahn Banach Theorem (in Addendum on problems), 194
Halmos, 129, 151–153, 157
heat equation, 3, 70, 118–119
Hellwig, 116
Herbrand, 78
higher order recursion theory, 192
Hilbert space [see Second Main Theorem, Eigenvector Theorem]
Hilbert space (definition), 8
Hilbert transform, 109

I

I^q [see computable rectangle]
I^q_M [see computable rectangle]
InEq (in proof of the Second Main Theorem), 178
injection operator, 107–108
Insertion Property, 81
integration, 12, 33, 35, 37–40
intrinsic computability [see computability]
isometries, 125, 145–148 [see also counter-examples]

K

Karp, 192
KdV equation, 194
Kirchhoff's formula, 12, 33, 73, 115–116
Kleene, 7, 25
Ko, 192
Kreisel, 2, 5, 13, 41, 193

L

L^p, 8 [see also computability]
ℓ^p, 8 [see also computability]
L^∞ [see computability, counterexamples]
ℓ^∞_0 [see computability]
Lachlan, 5
Lacombe, 4, 5, 12–13, 25, 28, 41, 104, 193 [see also computability]
Laplace's equation, 3, 70, 119
Limit Axiom [see axioms]
Linear Forms Axiom [see axioms]

linear independence [see Effective Independence Lemma]
linear span (definition), 78
Loomis, 151–153

M

$[-M, M]$ (in proof of the Second Main Theorem), 160
Markov, 78, 192
Mazur, 5, 17 [see also computability, Banach/Mazur]
Metakides, 5, 194
modulus of convergence, 16
monotone convergence, 20
Moschovakis, 5
Mostowski, 5, 24
Mycielski, 28
Myhill, 5, 50–53, 105

N

Nagy [see Riesz/Nagy]
Navier–Stokes equation, 194
Nerode, 5, 194
non normal operators, 194 [see also counterexamples]
nonlinear analysis, 194
Norm Axiom [see axioms]
normal operators, 123, 126, 132–133, 157, 184–187 [see also Second Main Theorem]
normal operators (definition), 126
Not an eigenvalue! (in proof of the Second Main Theorem), 168–169, 176–177

O

operational calculus (for the Spectral Theorem), 152–153, 161–162
operators [see bounded, unbounded, closed, self-adjoint, normal, compact; see also effectively determined operator]
orthonormal basis, 136–137, 140–141
overlapping intervals, 149, 162, 166–167, 174

P

$P = NP$ problem, 192
pairing function (definition), 7
partial derivatives, 58, 72
Petrovskii, 115
piecewise linear functions, 83, 112

polynomial space, 192
polynomial time, 192
Post, 78
potential equation [see Laplace's equation]
Pour-El, 50, 64–65, 68, 105–106, 111, 118–120, 194 [see also Caldwell]
primitive recursive functions, 192
problems, 192–194

Q

quantum mechanics, 2, 124, 126–127

R

real analytic functions, 64
real numbers [see computability]
recursive function (description in terms of Turning machines), 6
recursive set (definition), 7
recursive topology, 193
recursively enumerable nonrecursive set, 6–7, 15, 22, 52, 56, 58, 62–63, 90, 102, 104, 113, 129–130, 135, 146, 189
recursively enumerable nonrecursive set (definition), 7
recursively enumerable set (definition), 6–7
recursively inseparable pair of sets, 7, 42, 65
recursively inseparable pair of sets (definition), 7
Rice, 5, 11–12, 16–17
Richards, [see Pour-El]
Riemann Mapping Theorem (in Addendum on problems), 193
Riemann surfaces, 193
Riesz Convexity Theorem (classical), 110
Riesz/Nagy, 125–129, 151–153, 157, 190
Robinson, 5, 17
Rogers, 5, 7

S

$\mathscr{S}$ [see axioms, computability structure]
Šanin, 192
Schwartz [see Dunford/Schwartz]
Second Main Theorem, 1–4, 77, 123–124, 128–130, 149–191
Second Main Theorem (statement), 128
see saw construction, 44
self-adjoint operators, 2, 123, 125–126, 128–132 [see also Second Main Theorem, Eigenvector Theorem]
self-adjoint operators (definition), 126
sequences [see axioms, computability, counterexamples]

Subject Index   205

Shepherdson, 5
Shohat/Tamarkin, 87
Shore, 5, 194
Shub [see Blum]
Simpson, 5
Smale [see Blum]
Soare, 7
Sobolev spaces [see computability]
spectral measure, 149, 151–152, 166, 170–172
Specker, 5, 11–13, 16, 41
Spectral Theorem (classical, noneffective), 149, 151–157, 170
spectrum, 123–124, 126–127 [see also Second Main Theorem]
spectrum (definition), 126
SpThm (in proof of the Second Main Theorem), 153–157, 172
standard functions, 27
step functions, 5, 79, 84–85, 112–114
subrecursive hierarchies, 192

T
Tamarkin [see Shohat/Tamarkin]
Taylor series, 60–61
Theorems [The redundant terms "computability/noncomputability" have largely been omitted in the indexed list which follows.]
— Analytic Continuation Theorem, 60
— Closure Under Effective Uniform Convergence, 34
— Compact Operators, 129
— Composition of Functions, 28–31
— Creation and Destruction of Eigenvalues, 130
— Differentiation Theorem for C^1, 51, 104
— Differentiation Theorem for C^2, 53
— Differentiation Theorem for the Sequence of n-th Derivatives, 55
— Effective Density Lemma, 86
— Effective Independence Lemma, 142
— Effective Modulus Lemma, 65
— Eigenvector Theorem, 133–134
— Entire Function Theorem, 62
— Expansion of Functions, 33
— Fejer's Theorem, 106
— First Main Theorem, 101
— Fourier Series (effective convergence of), 105
— Fourier Series and Transforms (L^p-computability of), 110
— Heat Equation, 119

— Integration Theorems, 35, 37–39, 104
— Intermediate Value Theorem, 41
— L^p-Computability for Varying p, 107
— Laplace's Equation, 119
— Max-Min Theorem, 40
— Mean Value Theorem, 44
— Non-Normal Operators (non-computable eigenvalues), 132
— Operator Norm, 129
— Patching Theorem, 32
— Plancherel Theorem, 111
— Potential Equation [see Laplace's Equation]
— Real Closed Field, 44
— Riemann-Lebesgue Lemma, 111
— Second Main Theorem, 128
— Second Main Theorem (converse parts), 189–190
— Second Main Theorem (for normal operators), 184
— Second Main Theorem (for unbounded operators), 188
— Sequence of Eigenvalues (non-computability of), 129
— Seqentially Computable but Not Computable Continuous Functions, 67
— Stability Lemma, 87
— Step Functions, 114
— Stieltjes, Hamburger, Carleman Theorem, 87
— Third Main Theorem [see Eigenvector Theorem]
— Translation Invariant Operators, 71
— Waiting Lemma, 15
— Wave Equation Theorem (energy norm), 118
— Wave Equation Theorem (uniform norm), 116
— Wave Equation Theorem (weak solutions), 73
— Weierstrass Approximation Theorem, 45, 86
— Wiener Tauberian Theorem, 87
translation, 69–70
translation invariant operators, 51, 69–73
triangle functions (in proof of the Second Main Theorem), 149, 162–165, 167
Turing, 78
Turing machine, 6

U
unbounded operators, 1–2, 93–94, 123, 157, 187–188 [see also First Main Theorem]

uniform convergence, 12, 18–19, 33–34, 37, 44–49, 56, 72, 86, 105–106, 162
Uniformity in the Exponent Lemma, 158–160

W
wave equation, 3, 51, 65, 68–70, 72–73, 115–118, 120 [*see also* counter-examples, Theorems]
weak solutions, 73

weak topologies, 99–100, 108, 116
well understood functions, 112–114

X
$\{x_n\}$ (in proof of the Second Main Theorem), 160

Z
Zgymund, 110